Takeaways from Teaching through a Pandemic

In this insightful volume, more than 50 educators from 4 continents outline thoughtful and intentional innovations of lasting value made in their teaching of tertiary mathematics and statistics, in response to COVID-19 pandemic-related campus closures. The examples given in 20 practical chapters fall into three themes: utilization of relevant technologies, discipline-appropriate assessment alternatives, and support for learning and engagement.

The first theme explored is the utilization of the affordances of communications technology and mathematical software and online tools, to create learning resources and activities for virtual classrooms. The second theme is the design of sound assessment alternatives, together with the associated issue of maintaining academic integrity, in disciplines accustomed to established question styles and face-to-face exams. Finally, a range of activities to encourage engagement and support learning and teaching, both online and as campuses re-opened, is described.

Due to their ongoing relevance and benefits for tertiary mathematics education, be that of pre-service teachers, engineers or mathematics majors, the examples found in these 20 practical chapters are 'takeaways' or 'keepers'. This book was originally published as a special issue of *International Journal of Mathematical Education in Science and Technology.*

Katherine Seaton (FAustMS, SFHEA) is currently Adjunct Associate Professor at La Trobe University, Australia, following twenty-seven years of mathematics teaching, outreach, service and research. Her publications range from statistical physics to tertiary mathematics assessment and academic integrity, and mathematical fibre art.

Birgit Loch (FAustMS, PFHEA) is currently Dean of the Faculty of Science, Agriculture, Business and Law at the University of New England, Australia. With a background in computational mathematics, her research expertise is in innovation and educational technologies in tertiary STEM education.

Elizabeth Lugosi is Professor at the Mathematics, Statistics, and Data Science Department of Montgomery College. She is passionate about using engagement strategies to help students succeed in mathematics courses using collaborative, active learning and evidence-based instructional practices to enhance students' problem-solving and to build community.

Takeaways from Teaching through a Pandemic

Practical Examples of Lasting Value in Tertiary Mathematics Education

Edited by
Katherine Seaton, Birgit Loch and Elizabeth Lugosi

Routledge
Taylor & Francis Group
LONDON AND NEW YORK

First published 2024
by Routledge
4 Park Square, Milton Park, Abingdon, Oxon OX14 4RN

and by Routledge
605 Third Avenue, New York, NY 10158

Routledge is an imprint of the Taylor & Francis Group, an informa business

British Library Cataloguing in Publication Data
A catalogue record for this book is available from the British Library

ISBN13: 978-1-032-62747-2 (hbk)
ISBN13: 978-1-032-62748-9 (pbk)
ISBN13: 978-1-032-62749-6 (ebk)

DOI: 10.4324/9781032627496

Typeset in Minion Pro
by Newgen Publishing UK

Publisher's Note
The publisher accepts responsibility for any inconsistencies that may have arisen during the conversion of this book from journal articles to book chapters, namely the inclusion of journal terminology.

Contents

Citation Information

The chapters in this book were originally published in the *International Journal of Mathematical Education in Science and Technology*, volume 53, issue 3 (2022). When citing this material, please use the original page numbering for each article, as follows:

Introduction

Takeaways from teaching through a global pandemic – practical examples of lasting value in tertiary mathematics education
Katherine Seaton, Birgit Loch and Elizabeth Lugosi
International Journal of Mathematical Education in Science and Technology, volume 53, issue 3 (2022), pp. 559–565

Chapter 1

Any advice? Lessons learned by mathematics lecturers for emergency remote teaching during the COVID-19 pandemic
Eabhnat Ní Fhloinn and Olivia Fitzmaurice
International Journal of Mathematical Education in Science and Technology, volume 53, issue 3 (2022), pp. 566–572

Chapter 2

Creative approaches to teaching mathematics education with online tools during COVID-19
Sharyn Livy, Tracey Muir, Carol Murphy and Allison Trimble
International Journal of Mathematical Education in Science and Technology, volume 53, issue 3 (2022), pp. 573–581

Chapter 3

Your online textbook is ready: a shareable, interactive online textbook in response to COVID-19 lockdowns
Peter K. Dunn, Elizabeth A. Brunton and Michael B. Farrar
International Journal of Mathematical Education in Science and Technology, volume 53, issue 3 (2022), pp. 582–593

Chapter 4
Stop-motion LEGO® animations for learning linear algebra
Emily Cook
International Journal of Mathematical Education in Science and Technology, volume 53, issue 3 (2022), pp. 594–602

Chapter 5
Crowdsourcing solutions in the online mathematics classroom
Lorelei Koss
International Journal of Mathematical Education in Science and Technology, volume 53, issue 3 (2022), pp. 603–609

Chapter 6
Leveraging the power of Google Apps to support active learning in a synchronous online environment
Patrick Sullivan
International Journal of Mathematical Education in Science and Technology, volume 53, issue 3 (2022), pp. 610–618

Chapter 7
Flexible, student-centred remote learning for programming skills development
Peter Rowlett and Alexander S. Corner
International Journal of Mathematical Education in Science and Technology, volume 53, issue 3 (2022), pp. 619–626

Chapter 8
Don't throw the student out with the bathwater: online assessment strategies your class won't hate
Stuart Johnson, John Maclean, Raymond F. Vozzo, Adrian Koerber and Melissa A. Humphries
International Journal of Mathematical Education in Science and Technology, volume 53, issue 3 (2022), pp. 627–638

Chapter 9
Forced to improve: open book and open internet assessment in vector calculus
Tracy S. Craig and Tugce Akkaya
International Journal of Mathematical Education in Science and Technology, volume 53, issue 3 (2022), pp. 639–646

Chapter 10
No exam: assessment of third-year engineering students on the basis of self-generated Statistics cases
Nelly Litvak and Fulya Kula
International Journal of Mathematical Education in Science and Technology, volume 53, issue 3 (2022), pp. 647–655

Chapter 11

Utilizing discussion boards for test questions: opportunities for students' mathematical creativity and uniqueness
Miloš Savić
International Journal of Mathematical Education in Science and Technology, volume 53, issue 3 (2022), pp. 656–661

Chapter 12

Mathematics assessment integrity during lockdown: experiences in running online un-invigilated exams
Steven Richardson
International Journal of Mathematical Education in Science and Technology, volume 53, issue 3 (2022), pp. 662–672

Chapter 13

Mathematics and academic integrity: institutional support at a Canadian college
Josh Seeland, Lynn Cliplef, Caitlin Munn and Craig Dedrick
International Journal of Mathematical Education in Science and Technology, volume 53, issue 3 (2022), pp. 673–680

Chapter 14

Individualized summative assessments as used during COVID-19
Nathan Clisby and Antony Edwards
International Journal of Mathematical Education in Science and Technology, volume 53, issue 3 (2022), pp. 681–688

Chapter 15

Learning at a distance: can at-home activities measure up?
Jessica Gehrtz, Raquel Vallines Mira, Christopher Duffer and Priya V. Prasad
International Journal of Mathematical Education in Science and Technology, volume 53, issue 3 (2022), pp. 689–697

Chapter 16

Peer-Assisted Learning in a pandemic
Jessica Hargreaves, Claire Ketnor, Ellen Marshall and Sue Russell
International Journal of Mathematical Education in Science and Technology, volume 53, issue 3 (2022), pp. 708–716

Chapter 17

Checkpoints for online mathematics students with learning challenges
James William Dalitz
International Journal of Mathematical Education in Science and Technology, volume 53, issue 3 (2022), pp. 717–727

Chapter 18

Crafting connections in post-COVID classrooms: learning university mathematics through craft
Jeanette C. McLeod, Phillip L. Wilson, David Pomeroy and Julie Alderton
International Journal of Mathematical Education in Science and Technology, volume 53, issue 3 (2022), pp. 728–737

Chapter 19

The pandemic as a catalyst for rethinking active learning practices in technology intensive instructional environments
David Christopher Webb
International Journal of Mathematical Education in Science and Technology, volume 53, issue 3 (2022), pp. 738–744

Chapter 20

The unexpected consequences of replacing corridor chats with an online community of practice
Anita L. Campbell, Tracy S. Craig, Batseba Mofolo-Mbokane and Pragashni Padayachee
International Journal of Mathematical Education in Science and Technology, volume 53, issue 3 (2022), pp. 745–751

Notes on Contributors

Tugce Akkaya is Lecturer of mathematics at the University of Twente, the Netherlands, and my primarily focus lies in the realm of the implementation of digitalization in mathematics education.

Julie Alderton is Associate Professor in mathematics education (Faculty of Education) at the University of Cambridge, UK.

Elizabeth A. Brunton is an ecologist with a passion for teaching science and applied research skills.

Anita L. Campbell is a passionate educator and researcher at the University of Cape Town, South Africa, specializing in engineering education and positive psychology, dedicated to supporting student success and thriving.

Lynn Cliplef is Faculty Development Coach at Assiniboine Community College, Canada.

Nathan Clisby is a mathematician at Swinburne University of Technology, Australia, with research interests in statistical mechanics, enumerative combinatorics, and climate science.

Emily Cook is Senior Lecturer in STEM Education, with research interests in maths anxiety, project-based learning and incorporating arts and making into STEM learning.

Alexander S. Corner is Senior Lecturer at Sheffield Hallam University, England, primarily teaching pure mathematics to mathematics and computing students, while pursuing research in category theory and mathematics pedagogy in higher education.

Tracy S. Craig is Lecturer of mathematics at the University of Twente, the Netherlands, and primarily interested in practice-led research into the teaching and learning of calculus and linear algebra.

James William Dalitz has a strong and diverse background in science and engineering and presently works as a numeracy learning advisor at a regional university in Far North Queensland, Australia.

Craig Dedrick is a learning strategist at Assiniboine Community College, Canada.

Christopher Duffer is Lecturer in the Department of Mathematics at the University of Texas at San Antonio, USA.

Peter K. Dunn is an award-winning teacher who has taught introductory statistics for over 30 years at tertiary level.

Antony Edwards (now Sowards), Associate Director in Educational Futures at Swinburne University, Australia, is a researcher in mathematics education with a particular interest in student behaviour in tertiary-level mathematics and mathematics support.

Michael B. Farrar is a data and environmental scientist with a passion for sustainability and education. He has taught tertiary statistics for over 10 years.

Olivia Fitzmaurice is Senior Lecturer of mathematics education and academic director of the mathematics learning centre at the University of Limerick, Ireland.

Jessica Gehrtz is Assistant Professor in the Department of Mathematics at the University of Texas at San Antonio, USA, with research focused on college STEM instructors' use of student thinking in their teaching.

Jessica Hargreaves is Lecturer in data science at the University of York, England.

Melissa A. Humphries is a statistician – in a world overflowing with data, she creates, and applies, analytical tools to build our understanding of what it all means. She is a lecturer at the University of Adelaide, Australia.

Stuart Johnson has over 20 years of experience teaching tertiary maths, largely in first year mathematics with a particular interest in utilization of technology in assessment design.

Claire Ketnor is Principal Lecturer in mathematics at Sheffield Hallam University, England.

Adrian Koerber has been the director of first year mathematics at the University of Adelaide, Australia for over a decade.

Lorelei Koss is Professor in the Department of Mathematics and Computer Science at Dickinson College, USA. In addition to her interest in teaching undergraduate mathematics, she enjoys research on complex dynamical systems and ergodic theory.

Fulya Kula is Lecturer in the Applied Mathematics Department at the University of Twente, the Netherlands.

Nelly Litvak is Professor in the Applied Mathematics Department at the University of Twente and Eindhoven University of Technology, the Netherlands.

Sharyn Livy is Senior Lecturer of mathematics teacher education in the School of Curriculum, Teaching and Inclusive Education, Faculty of Education, Monash University, Australia.

Birgit Loch (FAustMS, PFHEA) is currently Dean of the Faculty of Science, Agriculture, Business and Law at the University of New England, Australia. With a background in computational mathematics, her research expertise is in innovation and educational technologies in tertiary STEM education.

Elizabeth Lugosi is Professor at the Mathematics, Statistics, and Data Science Department of Montgomery College, USA. She is passionate about using engagement strategies to help students succeed in mathematics courses using collaborative, active learning and evidence-based instructional practices to enhance students' problem-solving and to build community.

John Maclean is Lecturer (mathematical sciences) at the University of Adelaide, Australia. He loves dogs, data science, differential equations, and alliteration.

Ellen Marshall is a statistics lecturer and statistics support coordinator, Sheffield Hallam University, England.

Jeanette C. McLeod is Senior Lecturer in pure mathematics at the University of Canterbury, New Zealand and Co-founder and Director of Maths Craft New Zealand.

Raquel Vallines Mira is Professor in the Department of Mathematics at the University of Texas at San Antonio, USA, and in the Education Department at Universidad Europea del Atlántico, Spain, with research interest in the development of knowledge for pre-service teachers and mathematics teacher educators.

Batseba Mofolo-Mbokane is Mathematics Education Lecturer at the University of the Witwatersrand, specializing in teacher training, with research focus on innovative teaching strategies and effective learning of Calculus and other methods that assist conceptual understanding.

Tracey Muir is Professor in education (STEM) at the Australian Catholic University.

Caitlin Munn is an education quality assurance specialist at Assiniboine Community College, Canada.

Carol Murphy is Senior Lecturer in mathematics education at the School of Education, University of Tasmania, Australia.

Eabhnat Ní Fhloinn is Associate Professor in the School of Mathematical Sciences in Dublin City University, Ireland and is the Director of DCU Maths Learning Centre, Ireland.

Pragashni Padayachee is Associate Professor of mathematics education at the University of Cape Town, South Africa, devoted to advancing the teaching and learning of university mathematics through research, with a focus on innovative pedagogical approaches and student-centred strategies.

David Pomeroy is Senior Lecturer at the School of Teacher Education, University of Canterbury, New Zealand.

Priya V. Prasad is Associate Professor of mathematics at the University of Texas at San Antonio, USA, who studies how teachers at all levels develop mathematical knowledge and take up new classroom practices.

Steven Richardson is Senior Lecturer in mathematics at Edith Cowan University, Australia.

Peter Rowlett is a Reader at Sheffield Hallam University, England, teaching across a range of topics including mathematical modelling, game theory, programming and history of mathematics, and conducting research into higher education teaching practice.

Sue Russell is Academic Skills Manager at the University of York, England.

Miloš Savić is Professor of mathematics at the University of Oklahoma, USA and Director for the Undergraduate Research and Creative Activities Office.

Katherine Seaton (FAustMS, SFHEA) is currently Adjunct Associate Professor at La Trobe University, Australia, following twenty-seven years of mathematics teaching, outreach, service and research. Her publications range from statistical physics to tertiary mathematics assessment and academic integrity, and mathematical fibre art.

Josh Seeland is manager of library services at Assiniboine Community College, Canada.

Patrick Sullivan is Associate Professor of mathematics at Missouri State University (Springfield, Missouri), USA.

Allison Trimble is a retired researcher in education.

Raymond F. Vozzo is an education specialist at the University of Adelaide, Australia, with over ten years' experience teaching at all levels with a strong focus on first year undergraduate courses.

David Christopher Webb is Associate Professor (School of Education) at the University of Colorado Boulder, USA. His research focuses on the design of STEM instructional environments that are more conducive to meaningful student engagement, persistence, and learning.

Phillip L. Wilson is Senior Lecturer in applied mathematics at the University of Canterbury, New Zealand and Co-founder and Deputy Director of Maths Craft New Zealand.

Introduction to takeaways from teaching through a global pandemic: practical examples of lasting value in tertiary mathematics education

Welcome to this Special Issue dedicated to what the tertiary mathematics education community around the world has learnt and developed through our rapid shift to online teaching during the global pandemic. All but one of the previous Special Issues of IJMEST have arisen out of conferences, and that other one related to a particular funded project. This Special Issue is thus unique, there having been an open call for Expressions of Interest to contribute to it, with the intention of collating and sharing valuable insights from responses to such an extended and unexpected worldwide event.

This Special Issue is also this journal's first Special Issue that consists solely of *Classroom Notes*. *Classroom Notes* are shorter articles highlighting a single, unusual, and interesting result, an alternative method, or a creative approach to teaching a particular mathematical idea. The *Classroom Notes* here are intended to provoke ideas that will support our teaching of mathematics at university level and focus on examples of practical and pedagogical use to readers. While not required to be research articles, they are nevertheless scholarly, have been double-blind peer-reviewed, comply with ethical principles and are of lasting value to the mathematics education community. We would like to take this opportunity to thank again the many reviewers without whose generosity of time and expertise such a large Special Issue could not have come into being.

The rapid move to 'emergency' online teaching in early 2020, caused by lockdowns across the world to tackle the spread of COVID-19 and protect lives, has disrupted mathematics education, and will have long-lasting effects on how we teach. It is likely that some form of online teaching and learning will remain, even after vaccines have been rolled out widely. It was exciting to see how mathematics educators have embraced the challenges posed by the global pandemic and have been very thoughtful and innovative in their responses. This Special Issue brings together valuable examples which go beyond being temporary stop-gap measures, to being of practical use in a post-pandemic future, new practices or insights which we have termed *takeaways*. Clearly the theme of the Special Issue struck a chord both with experienced authors in Scholarship of Teaching and Learning, and with others new to publishing in this area. We were gratified to receive more than 60 proposals, from around the world and across most of the sub-themes we originally identified in the call for Expressions of Interest. However, that did mean that we had to make some tough decisions to reflect a truly international flavour as well as a balance of content in the 20 articles that we ultimately included. Because they support the theme of this Special Issue, we have included several *Classroom Notes* which summarize actions and responses from multiple institutions or which describe professional support developed by and for tertiary mathematics educators during 2020 – topics that would not usually fall within the scope of a *Classroom Note*.

In one such piece, Ní Fhloinn and Fitzmaurice provide a good launching place as we introduce the included papers. We expect many readers to nod in agreement when they read this *Classroom Note*, particularly with respect to the wide range of useful advice provided. Ní Fhloinn and Fitzmaurice present an analysis of thoughts and advice collected from 257 lecturers in 29 countries (mostly European), just two months after the start of the pandemic, as part of an overarching study in which the two authors are involved. Ní Fhloinn and Fitzmaurice touch on four themes: technology options, online teaching approaches, supporting students, and setting realistic expectations for oneself. They found no agreement among their respondents as to what software is most suitable for teaching mathematics remotely, with an abundance of tools mentioned, reflecting the willingness of lecturers to innovate and try new approaches rather than embrace what is provided through their institutions. There was also no consensus on whether asynchronous, synchronous or mixed approaches work best. Perhaps not surprisingly, seeking real-time feedback from students on new online approaches is one suggestion; another is to structure their study and articulate expectations more explicitly than might have been the practice in face-to-face teaching. The use of pre-existing material as much as possible is proposed, so more time can be spent focusing on assessing student understanding and fostering engagement, while letting go of perfectionism when producing resources in difficult times is encouraged.

The remaining *Classroom Notes* fall into three strands: adoption of technology, assessment and academic integrity, and engagement and support of learning.

It is almost impossible to imagine how education could have continued during lockdowns and campus closures if it were not for the affordances of current technology, and it goes almost without saying that the teaching of university mathematics in the pandemic required its incorporation. Upskilling in the use of technology by both educators and students, and the teaching innovations it has led to, would have come far more slowly without the (regrettable) impetus of the pandemic. Livy, Muir, Murphy and Trimble, who educate pre-service primary (elementary) mathematics teachers in two Australian universities, introduce us to a helpful framework – the SAMR (Substitution, Augmentation, Modification and Redefinition) model – for viewing the use of technology in the online setting and for taking it into a post-Covid future. They critique their own practice, noting that intended substitution (e.g. videos replacing face-to-face lectures) can be in fact augmentation (because students can pause or rewatch). An example they give of modification is the use of tools permitting anonymous (and thus less hesitant) contributions to discussions, while an entirely new modelling activity made possible by using the online tool *Desmos* exemplifies redefinition. We would encourage readers to apply this framework to reflect on and evaluate their own use of technology in teaching mathematics during 2020 and 2021, and to consider what transformations they can incorporate into future settings.

Dunn, Brunton and Farrar provide detailed, practical instructions and advice so that readers may follow their steps to create a bespoke online textbook using freely available software. They discuss the pedagogical benefits of being able to embed interactive digital tools, the ease of keeping such a text current, the lasting access that students have to such a book compared to LMS (Learning Management System) resources, and its affordability, compared to a traditional text. Their book, developed for a first-year large-enrolment statistics subject taken by students from a wide range of disciplines in an Australian university, is available as an open educational resource (OER), quite literally a 'takeaway'. Developed during emergency remote teaching, the book is now incorporated into their face-to-face delivery of the same subject.

Cook used the time she was forced to spend at home during extended lockdowns in Melbourne to develop short LEGO® stop-motion videos explaining linear algebra concepts, processes and problems in a fun and humorous way, albeit underpinned by relevant scholarship. These videos are another literal takeaway, being available on YouTube, and they are well worth watching, both for their educational and their entertainment contributions. Cook acknowledges that producing these videos is a significant time commitment; to take this idea into the future the work and the product could be shared, with a library of videos built up over time. Though developed to replace the gestures and spontaneous diagrams of live lectures during remote teaching, the videos will be used in the future during lectures and as resources in the LMS.

Koss describes her attempt at a 'crowdsourcing' approach to solving problems involving graphing or visualization in vector calculus and complex analysis. This approach met with mixed success in face-to-face teaching pre-pandemic, but it proved to benefit from the technology-based approach required in online classes. Students could draw synchronously on the shared whiteboard tool in Zoom breakout rooms, quickly and with the possibility to fix mistakes as they went. Small components were being developed in parallel to give the solution of a larger problem, although Koss identifies that such a solution could be developed by students working asynchronously, too. One unexpected takeaway to consider is that simultaneous drawing using technology could be enacted in the face-to-face environment in the future using conferencing technology such as Zoom, even with participants in the same physical classroom.

An essential goal of Sullivan, who like Koss is US-based, was to create active group collaboration in a synchronous online learning environment, as he had in his previous face-to-face classes. To minimize the disruption that can occur with an instructor entering a breakout room, the author established an online environment for group work using multiple Google apps, wherein students could work uninterruptedly with peers while receiving immediate written feedback or prompts from the teacher, who could simultaneously monitor all groups. Sullivan anticipates that this strategy will be beneficial, and in some ways superior, for in-person classroom discussions in the future.

Rowlett and Corner reimagined the teaching of programming to mathematics students at a UK university, providing neither recorded instruction nor synchronous online classes as substitutes for the pre-Covid student-centered tutorials. Rather, the time of the module-delivery staff was utilized to supply frequent formal formative feedback on tasks, thus creating prompts for students to engage with the help available by means such as email, live text chat, video calls and a scheduled online 'drop-in'. They observe (as a takeaway) that this formal process was effective in identifying misconceptions or problems with code which may be missed, by student and by teacher, in the informal interactions of face-to-face classes. They also touch on an equity issue that does need to be considered; not all students are able to join extended synchronous online classes, for a variety of reasons. It can be possible to teach some topics, such as programming, without requiring this.

Like a significant aftershock to the seismic shift of moving all teaching online, planning for valid end-of-semester assessment soon became a pressing concern. Before the pandemic, there was little that had been published regarding academic integrity in mathematics specifically, and the use of face-to-face invigilated final exams was widespread. In this prior sense, 'open book' meant quite literally just a bound volume; in the online environment it could mean 'open web'. Computer-aided exams, where they existed, took place in a controlled environment, not at all like the situation in a student's home. Limiting access to calculators, be they scientific or graphical or computer-algebra enabled, had been a choice that educators could impose as they saw fit. In writing questions for an online exam, free access to such tools needed to be considered. Could, or should, an exam be replaced by another form of assessment? A third of the *Classroom Notes* included in this Special Issue describe approaches intended to provide valid assessment and to protect against academic integrity breaches (or, failing that, to identify them when they occur). As Craig and Akkaya write, 'the coronavirus situation has forced many of us to change our assessment practices in creative ways which, while uninvited, have proved to be worthwhile and worth maintaining.'

Maclean, Humphries, Johnson, Koerber and Vozzo discuss two case studies which highlight different approaches to nonintrusive online assessment taken within the same university department in Australia. In the first case study, set in the context of a large first-year mathematics course, intentional question re-design was used to reduce the usefulness of computer algebra tools and the potential for outsourcing in an open resource exam, addressing their stated primary concern that if the challenge could be offloaded, the potential for them as assessors to differentiate between levels of student skills would be eroded. The second case study, set in a first-year statistics course, moved away from a single high-weight exam entirely, replacing it with a suite of tasks, including group project work throughout the semester, supplemented by a timed individual task relating directly to this project. The authors identify, as takeaways for their readers and their own future practice, that steps as simple as modified question design and increased use of continuous assessment can be effective to achieve differentiation between students in the changed environment.

Craig and Akkaya evaluate their experience of moving assessment online in positive terms as being 'forced to improve'. When assessing students in a vector calculus course, taken by second-year engineering and physics students in a university in the Netherlands, for fairness (since some students might think to access technological tools and others not), they downplayed the computational aspects, instead placing the specific higher-order skill of informed judgement in decision making at the heart of what was assessed. For those readers who do not teach vector calculus, they point out other content areas where informed judgement is also a good sign of student learning, and their enthusiasm for the benefits of being shaken from the rut of procedural questions is apparent.

From the same university, Kula and Litvak describe assessment of statistics for mechanical engineering students without any exam at all. Rather, providing freedom within boundaries, they required their students to find their own unique case studies from an engineering context, formulate clear statistical questions, solve these questions, and identify the learning objectives they had addressed. This shifted the focus from calculations to choosing an appropriate statistical test, and reflection. This *Classroom Note* also makes the point, as do some others, that assessment changes must be supported with practice in class. The takeaway here comes with the caveat that a no-exam assessment regime must be aligned to the whole curriculum delivery.

Savić is also positive about his experience of replacing customary pen-and-paper assessments in a small-class context in a university in the US. He found himself able to fully realize his long-held beliefs about learning and problem solving – that it should be collaborative and that it requires what he terms 'incubation' time to solve a problem creatively, neither of them features offered by traditional timed tests. He posted a generally-stated problem on an electronic discussion board, and each student was asked to propose their own unique solution, with plenty of time (up to a week) and one another's responses visible to them. He thus fostered creativity, thoughtfulness, and collaboration without test anxiety. In his own words 'Creating discussion boards that showcase students' uniqueness while demonstrating understanding of content will be in my pedagogical toolbox for the rest of my career.'

Back in an Australian large-class context, Richardson discusses several challenges. As is done in some other *Classroom Notes* too, he describes using the affordances of software to randomize or to individualize questions by varying parameters in one part of the exam, with a second written part to be uploaded showing reasoning and some reflective writing. He also discusses, with examples, appropriate and effective question design for the online situation. This is a frank paper, in which we can also read about things that did not work as well as he had hoped, and the frustrations for the assessor of student academic misconduct and the associated investigative processes. The identified takeaways are recommendations both for exams themselves and for the processes used to investigate misconduct; in particular the need for people with relevant expertise to be involved, since non-experts do not always understand what misconduct in mathematical tasks looks like.

By contrast, Seeland, Cliplef, Munn and Dedrick, writing from a small Canadian college, demonstrate the willingness of non-experts in mathematics to consider and accommodate the nature of mathematical work in a whole-of-institution approach to academic integrity. They eschew a reactive, negative response, adapting and leveraging positive and educative approaches shown to be effective in text-based disciplines. The societal good of learning with integrity, such as becoming a safe nurse or engineer, supplants the mere mechanics of integrity by giving it purpose. They also emphasize the role of adequate academic support, in their institution provided centrally in consultation with the discipline, to reduce anxiety and build self-efficacy, reducing the temptation to cheat.

The last assessment-focused *Classroom Note*, by Clisby and Edwards, explores the theme of assessment and academic integrity. In addition to describing individualization, they conceive a helpful model for viewing online examination design, delivery, and feedback as distinct stages. From their experience within an Australian mathematics and statistics department, they describe a 'platform-agnostic' way in which these ideas can transfer to others, each of whom can use the most appropriate software, or the one with which they are most comfortable, at each of these separate stages. Another useful takeaway from this *Classroom Note* is the use of Google alerts to identify from key esoteric phrases when contract cheating services are being sought for a task; both the pre-emptive timing and the automation of this strategy are sure to be of interest to many readers.

The final cluster of papers that we introduce are those that address the issues of equity, fostering of engagement, and provision of support when in-person interactions have been restricted.

Gehrtz, Vallines Mira, Duffer and Prasad describe how an introductory task redesigned to be undertaken at home formed part of their online teaching. Exploratory measurement activities allowed their students, prospective elementary teachers studying at a Hispanic-Serving Institution in the southwestern United States, to be better prepared for class, enabling more equitable participation in classwork and giving rise to richer discussions. The authors will continue to use this activity when returning to the in-person classroom and encourage other instructors to develop exploratory activities that students can do at home, making students better prepared to participate in a new topic in class.

Hargreaves, Cornock, Marshall, and Russell summarize for us their experiences in two UK university mathematics departments with maintaining Peer-Assisted Learning (PAL) schemes during the pandemic, wherein student volunteers design and deliver sessions supporting student groups in lower years. They observed that remote PAL was valuable because of its accessibility and inclusivity, providing added anonymity and flexibility for those seeking assistance. One key takeaway from this time for their PAL schemes and wider teaching is a new realization of the potential of online tools which can be integrated into face-to-face settings by all involved in mathematics teaching and learning.

Dalitz presents an approach adopted with his Australian students, which he terms Checkpoints, used in the virtual environment to support students with learning challenges, who are under-prepared, or are affected by mathematics anxiety. His main goal was to provide the same quality of support as was possible through interpersonal interactions in face-to-face circumstances. The Checkpoints contained signposts to support student learning of content, or tips for wellbeing, learning, or assessment. The Checkpoints regularly provided opportunities to reflect on learning, and reassurance of being on the right track. The author considers such Checkpoints useful in any type of class, and beneficial for all students regardless of their affinities towards mathematics. He advises writing Checkpoints with care; they need to be supportive, not give extra work, and provide benefits for students other than merely including additional mathematics content.

McLeod, Wilson, Pomeroy and Alderton describe re-engaging final year University mathematics students with concepts and with each other through hands-on mathematical craft. These authors demonstrated the effectiveness of visualization to help students understand non-trivial mathematical concepts. Survey results showed that the construction of mathematical craft objects, an activity introduced when students returned to live classes after the lockdown in New Zealand, played a significant role in an enhanced sense of community, reduced learning-related stress, and increased engagement with learning mathematics.

Turning to support for educators themselves, Webb shares how the Student Engagement in Mathematics through an Institutional Network for Active Learning (SEMINAL) project, started in 2018, shifted focus and modus operandi at the onset of the pandemic to support colleagues from twenty-three university mathematics departments in the United States. The SEMINAL Networked Improvement Community (NIC) supported participants in adapting instructional practices for virtual environments to sustain a level of active learning, thus enabling student interaction under remote circumstances. During unexpected challenges, the SEMINAL NIC proved invaluable to address common problems and share resources.

Campbell, Craig, Mofolo-Mbokane and Padayachee experienced 'the unexpected consequences of replacing corridor chats with an online community of practice'. They have each been advocates of collaboration among students, emphasizing the benefits of brainstorming, understanding different perspectives, and challenging and critiquing one another. The pandemic provided them the incentive to create an online opportunity for support and collaboration for staff across two continents (Africa and Europe). Their Community of Practice involved regular meetings, homework readings, and supportive interrogation of one another's practice, resulting in new teaching and assessment alternatives they applied in their classes and shared with other colleagues. Two key ingredients to the success of this approach were a shared belief in student-centered values as a driving force for decisions during the pandemic, and the trust they had towards each other, providing a safety net to accept critiques from one another.

These last two *Classroom Notes* remind us that there will always be challenges in making teaching relevant to a changing world and hence potential value in finding support from wider colleagues.

This Special Issue explores through 20 *Classroom Notes* how mathematics educators around the world overcame the challenges posed by the COVID-19 emergency transition to remote learning. Authors of this issue did not shrink from the difficulties of online/blended teaching and learning; instead, they accepted the challenge, found motivation in it, and produced fantastic ideas for the benefit of their students, their professional development, and tertiary mathematics education. We are delighted to share these ideas with you.

Katherine Seaton
http://orcid.org/0000-0002-7354-8814
Birgit Loch
http://orcid.org/0000-0001-6497-9447
Elizabeth Lugosi
http://orcid.org/0000-0002-1818-8326

OPEN ACCESS

Any advice? Lessons learned by mathematics lecturers for emergency remote teaching during the COVID-19 pandemic

Eabhnat Ní Fhloinn and Olivia Fitzmaurice

ABSTRACT
The COVID-19 pandemic meant a move to emergency remote teaching for many university students around March 2020. Two months later, we undertook an online survey to ascertain experiences of university mathematics lecturers. We received responses from 257 mathematics lecturers from 29 countries. In this paper, we consider their responses when asked what advice they would give to colleagues, based on their experience of online teaching to date. We gather their collective wisdom under a number of themes, including technology options, effective online teaching approaches, and supporting students, as well as reducing stress and setting realistic expectations for both learners and teachers.

1. Introduction

The move to emergency remote teaching due to the COVID-19 pandemic necessitated a swift, unprecedented change in teaching delivery for universities globally. This paper reports on one part of a larger study into the remote lecturing of mathematics during the initial months of the pandemic. An anonymous survey investigating how mathematics lecturers were coping with teaching remotely was created, piloted and amended based on feedback from a panel of mathematics lecturers, and ethical approval was attained from Dublin City University. The survey was created in Google Forms and distributed in May 2020. We received responses from 257 lecturers based in 29 countries, the majority in Europe. Respondents were at different stages of their semester depending on where they were based. For full details of the study's background, see Ní Fhloinn and Fitzmaurice (2021). Here, we report on the replies given when respondents were asked what advice they would give to colleagues, based on their experiences of remote teaching. We have grouped their comments under the four most common themes: technology options; online teaching approaches; supporting students; and reducing stress and setting realistic expectations.

2. Technology options

While the choice of technology used was not always up to the lecturers, depending on cost, licences and availability, several hardware options were commonly recommended. In addition to the standard laptop and webcam, the most popular options were a visualizer,[1] and a tablet/graphics tablet with a stylus pen.

Although visualizer usage decreased from a quarter of respondents before the pandemic to 9% during emergency remote teaching, due to lack of access to campus facilities, those that did manage to use one recommended it highly. One respondent stated 'I used my visualiser hooked up to Panopto and the experience was seamless via Microsoft Teams', but advised doing separate recordings afterwards 'to have more precise recordings without interruptions by students'. Another respondent cautioned that 'recording via visualiser produces very large.avi files which couldn't be uploaded to Moodle'. For those unable to source or afford a visualiser, one respondent recommended that 'a smartphone is great as a visualiser, especially for recording and working through maths equations/problems'. This can be achieved in many ways, including downloading a purpose-built app, such as iDocCam or Reflector, or directly via Blackboard Collaborate (https://lta.hw.ac.uk/wp-content/uploads/14_Using-a-mobile-device-as-a-Visualiser_Document-camera.pdf), Teams (https://www.maths.ox.ac.uk/members/teaching-staff/teaching-remotely/android-visualiser), or Zoom (https://www.northwestern.edu/academic-support-learning/documents/how-to-use-your-cell-phone-as-a-document-camera-in-zoom.pdf).

Usage of iPads, tablets or graphic tablets increased during emergency remote teaching, rising from 13% of respondents prior to the pandemic to almost 40%. In general, these were highly recommended by respondents, with one observing that they 'prefer graphic tablet to visualizer because no lighting or focus issues and hand is not obscuring writing'. Respondents highlighted a number of different approaches for using these, such as via Teams ('Used MS Teams for Q + A sessions with students and used Wacom tablet and PowerPoint slides… to answer questions live via the shared screen facility') or Zoom ('I use my iPad on zoom… I write on the GoodNotes App, which auto-backups. The students have access to that file at any time so that they can scroll back up').

Advice in relation to software was far more varied, with respondents mentioning their own personal favourite but with little consensus among individuals (e.g. 'Webwork is the bees-knees', 'Discord is good', 'Explain Everything is great', 'PDF Annotator for writing by hand on a Surface computer is not a blackboard, but not too far from it'). In total, 109 different software packages were mentioned throughout the survey. Recommendations were often made based upon software that easily integrated with that already in use within their own university.

Teaching and learning mathematics online is recognized as challenging for teachers (Engelbrecht & Harding, 2005; Radmer & Goodchild, 2021). Several respondents acknowledged the steep learning curve involved in using new technologies, with one recommending 'Keep the technology as simple as possible. Remember the students have to adapt to this new way of doing things too.' The general advice was to 'attend as many training sessions as possible' and to 'experiment to find the right fit'.

3. Online teaching approach

A number of suggestions were made as to how to increase effectiveness while teaching in this environment. Advice was mixed in terms of live online sessions versus pre-recording material, with some advising 'don't do live lectures – too many things can go wrong', others recommending 'live teaching with interactions is absolutely essential' and more suggesting to 'combine live and non-live sessions'. There was greater consensus around the importance of ensuring easy interaction with students, with suggestions ranging from using 'online chat/ discussion forum for posting any time', to streaming on YouTube, where 'students can comment and ask questions on email or chat'.

The increased time needed to develop online teaching resources was cited numerous times, with lecturers advised to 'allow more time for preparation' and not to 'underestimate how time consuming it can be to put together something that's reasonably OK'. The choice between using pre-existing material or generating something new had respondents firmly in favour of the former, advising that lecturers should 'use existing resources (youtube videos, geogebra activities etc.) as much as possible and focus less on content generation and more on assessing student understanding and engagement', highlighting that 'material already exists that students can study and this material is far better than we produce in a limited time frame. Our job is to support students in learning that material.' Recent research in Ireland uncovered a preference for video resources when given the choice between attending mathematics lectures and/or watching videos online (Howard et al., 2018). The use of videos has been demonstrated to increase student motivation towards mathematics (Kinnari-Korpela, 2015). From a student's perspective, videos are more convenient and allow them to dictate their own pace (Howard et al., 2018; Trenholm et al., 2012).

On a practical level, one respondent advised that you should 'always appoint two class members to be your canaries-in-the-mine and give them the responsibility of providing you with real-time feedback if connection/software problems arise that affect the class' ability to read/hear/see anything.' Research shows that feedback from students to lecturers can have a positive impact on teaching and lead to improvements in delivery (Flodén, 2017; Keane & Labhrainn, 2005; Seldin, 1997).

Additional practical tips included designing your teaching bearing in mind what your students can access ('survey your students (anonymously) to find out what facilities they have, and use this to design what you produce') and in particular to 'prepare your material in bite-size chunks, in a format that ALL your students can access' in order to 'keep video file sizes relatively small'. Given that poor internet speed emerged as a major concern for respondents, both as lecturers and for their students, advice of this nature is a significant takeaway. It is supported by Hyland and O'Shea's (2021) survey of 263 undergraduate mathematics students in Ireland, where more than a third of respondents did not have access to efficient broadband or own a PC/laptop. This appears to be a global issue with similar findings reported in Australia (Dodd et al., 2021), Malaysia (Nassr et al., 2020), and the USA (Soria & Horgos, 2020).

Overall, there was concern about online assessment approaches, with several commenting that they worried about 'loss of integrity' within the assessment system, and that colleagues needed to 'think about how assessment will need to change'. That said, respondents were generally positive about the possibilities of online formative assessments, both during live sessions using polls ('ask lots of questions or run short polls when giving an online lecture... to keep the students involved and get as much feedback as possible') and using multiple-choice or computer-aided assessment for low-weighted assessments, often embedded into their institutional Virtual Learning Environment[2] (VLE) ('With the Moodle exam, [there are] quite good options (not just simple multi-choice questions)'). Managing assessment overload was another tip from respondents ('I have noticed a tremendous student overload with projects, assignments, and tests'), along with ensuring that students are prepared for the types of assessments they might face ('Our usual methods of teaching prepare students for our usual exams. If there's a risk of alternate exams due to another wave of infections then it would be better if the students were prepared for either type of exam').

The importance of providing feedback to students in an online environment was also the subject of several comments ('I would advise all educators to remember the value of feedback and try to hold Q and A sessions to make sure students understand'). Some suggested achieving this by 'using pre-recorded segments, and using lecture time for discussion, answering questions' while others advised 'acknowledging the work that the students have completed and providing feedback so that students can improve' through the use of homework or regular assignments. Provision of prompt feedback has long been at the centre of effective teaching and learning (Chickering & Gamson, 1987). Feedback engages and motivates students and helps bridge some of the distance that occurs with online learning (Tanis, 2020).

Several respondents highlighted the opportunity inherent in such a dramatic shift in teaching delivery, noting that 'this is a very good time to think about "starting anew" rather than just add on online increments' and that lecturers should not 'just try to replicate face-to-face teaching'.

4. Supporting students

Several commentators stressed the importance of interacting regularly with students to ensure engagement ('a variety of modes of engagement is important for the inclusion of students' differing learning styles and personalities') and provide support ('accept that some students will need a lot more emotional support'). One suggested designing teaching to make sure to give 'students an opportunity to still create bonds with their peers, as a lot of campus learning is by peer instruction'. Communication with students was highlighted as essential, so that they know 'what and how you want them to study each week', because 'as teachers, we say, tell them, tell them, and tell them again; you need to add about 5 "tell thems" to that and put it on all forms of communication: email/VLE announcement etc., keeping it brief and clear.'

Online teaching tends to lead to a reduction in interaction between lecturers and students and between students themselves (Trenholm et al., 2012). Radmer and Goodchild (2021) found the absence of live interaction with lecturers was missed by the majority of mathematics students they surveyed during the Covid-19 pandemic, and that interaction is important to these students. In fact, Tanis (2020) reported that interaction with lecturers is more valued than interaction with peers or content when learning online. Regular communication between lecturers and students can help support learning and alleviate student anxiety (Radmer & Goodchild, 2021).

Respondents also noted that 'online teaching is very difficult for first-year students as it needs a lot of self-discipline and ability to learn alone' and so, lecturers should ensure to 'clearly explain our expectations and provide a structure that students can use to achieve these expectations', with one advising 'do not say that "I have uploaded the material, read it."' Ensuring the student voice is heard was also recommended, by 'running suggestions through student representatives, if possible, as they also provided very useful viewpoints.'

5. Reducing stress and setting realistic expectations

Strikingly, many respondents gave advice on reducing stress and setting realistic expectations. Numerous comments were made along the lines of doing the best that you can ('Do your best, but don't hold yourself to impossible standards, and don't let others hold you to them either'), with one respondent stating that 'the students are not expecting "best practice" and appreciate it if you do your best'. The issue of perfectionism in creating online resources was raised, with respondents advising that 'if you trip over your words in a lecture, you correct yourself and carry on; you don't start again at the beginning' and that it is important to remember that a professional finish to any recording requires 'a lot of hardware, software, support and a LOT of time to spend on it'. One respondent advised 'not being too adventurous initially with unfamiliar tools – try to keep to the familiar', suggesting that maintaining your usual lecturing style in an online environment (using a physical whiteboard and video camera, or a graphic tablet) reduces some of the stress. Another highlighted that 'we have been doing emergency online teaching in a global pandemic, not preparing professional online courses'. Finally, several advised to avoid overworking, saying 'you're not going to develop fully fledged online courses for all modules in the space of one year. Way too much work. Be realistic and set realisable goals'. It is unsurprising that there was so much advice of this nature, given that 72% of respondents found online teaching to be stressful or very stressful initially, and 47% still found it thus when completing the survey.

6. Takeaways

The collated wisdom provided by 257 lecturers based in 29 countries during emergency remote teaching has been presented here. The findings suggest to us that a blend of old and new skills should be utilized moving forward. While some focused on specific technology-based approaches that worked well for them, one observed that 'there's a lot of noise around which is the best software/hardware to use and there is no correct answer. Just keep trying new things until you find the right combination for you'. Particular features that were valued included tools which required little prior training before usage, resulted in small file sizes, interacted well with tools already in use, and allowed for mathematics to be 'written' online. There was also an acknowledgement of how this online teaching experience could influence future teaching styles, as 'there is a vast amount of opportunities to incorporate online teaching into regular classes'. The creation of electronic resources that students can review at their own pace or use for formative assessment purposes can be a valuable reserve to support the teaching and learning of mathematics, and this is further explored in Fitzmaurice and Ní Fhloinn (2021). In terms of student support, consideration of student access to internet and hardware is crucial. These resources, married with frequent interaction, open lines of communication with lecturers and bi-directional feedback may provide the foundation for a new era of learning mathematics. Finally, although no longer in the early months of emergency remote teaching, the advice to set realistic goals within the context of moving to a new style of teaching is valuable to bear in mind, to ensure greater success in both the teaching and learning of mathematics in the future.

Disclosure statement

No potential conflict of interest was reported by the authors.

Notes

1. A visualiser is a mounted camera that projects an image of what a lecturer is writing/demonstrating onto a smart board for their class to see, sometimes called a document camera.
2. A Virtual Learning Environment provides a platform for supporting teaching, learning and assessment by acting as a content repository.

ORCID

Eabhnat Ní Fhloinn http://orcid.org/0000-0002-6051-2249

References

Chickering, A. W., & Gamson, Z. F. (1987). Seven principles for good practice in undergraduate education. *AAHE Bulletin*, 3–7. https://files.eric.ed.gov/fulltext/ED282491.pdf

Dodd, R. H., Dadaczynski, K., Okan, O., McCaffery, K. J., & Pickles, K. (2021). Psychological wellbeing and academic experience of university students in Australia during COVID-19. *International Journal of Environmental Research and Public Health*, *18*(3), 866. https://doi.org/10.3390/ijerph18030866

Engelbrecht, J., & Harding, A. (2005). Teaching undergraduate mathematics on the Internet. *Educational Studies in Mathematics*, *58*(2), 253–276. https://doi.org/10.1007/s10649-005-6457-2

Fitzmaurice, O., & Ní Fhloinn, E. (2021). Alternative mathematics assessment during university closures due to Covid-19. *Irish Educational Studies*. https://doi.org/10.1080/03323315.2021.1916556

Flodén, J. (2017). The impact of student feedback on teaching in higher education. *Assessment & Evaluation in Higher Education*, *42*(7), 1054–1068. https://doi.org/10.1080/02602938.2016.1224997

Howard, E., Meehan, M., & Parnell, A. (2018). Live lectures or online videos: Students' resource choices in a first-year university mathematics module. *International Journal of Mathematical Education in Science and Technology*, *49*(4), 530–553. https://doi.org/10.1080/0020739X.2017.1387943

Hyland, D., & O'Shea, A. (2021). The student perspective on teaching and assessment during initial COVID-19 related closures at Irish universities: Implications for the future. *Teaching Mathematics and its Applications: An International Journal of the IMA*. https://doi.org/10.15388/infedu.2015.05

Keane, E., & Labhrainn, I. (2005). *Obtaining student feedback on teaching & course quality*. Centre for Excellence in Learning and Teaching (CELT): NUI Galway. http://www.nuigalway.ie/media/celt/files/coursedesign/ReviewofTeachingEvaluationMethods.pdf.

Kinnari-Korpela, H. (2015). Using short video lectures to enhance mathematics learning-experiences on differential and integral calculus course for engineering students. *Informatics in Education-An International Journal*, *14*(1), 69–83. https://doi.org/10.15388/infedu.2015.05

Nassr, R. M., Aborujilah, A., Aldossary, D. A., & Aldossary, A. A. A. (2020). Understanding education difficulty during COVID-19 lockdown: Reports on Malaysian university students' experience. *IEEE Access*, *8*, 186939–186950. https://doi.org/10.1109/ACCESS.2020.3029967

Ní Fhloinn, E., & Fitzmaurice, O. (2021). Challenges and opportunities: Experiences of mathematics lecturers engaged in emergency remote teaching during the COVID-19 pandemic. *Mathematics*, *9*(18), 2303. https://doi.org/10.3390/math9182303

Radmer, F., & Goodchild, S. (2021). Online mathematics teaching and learning during the COVID-19 pandemic: The perspective of lecturers and students. *Nordic Journal of STEM Education*, *5*(1). https://doi.org/10.5324/njsteme.v5i1.3914

Seldin, P. (1997). Using student feedback to improve teaching. *To Improve the Academy*, *16*(1), 335–345. https://doi.org/10.1002/j.2334-4822.1997.tb00333.x

Soria, K. M., & Horgos, B. (2020). *Social class differences in students' experiences during the COVID-19 pandemic*. SERU Consortium Reports. Center for Studies in Higher Education: UC Berkeley. Retrieved March 29, 2020, from https://escholarship.org/uc/item/3hw2m00g

Tanis, C. J. (2020). The seven principles of online learning: Feedback from faculty and alumni on its importance for teaching and learning. *Research in Learning Technology*, 28(0). https://doi.org/10.25304/rlt.v28.2319

Trenholm, S., Alcock, L., & Robinson, C. L. (2012). Mathematics lecturing in the digital age. *International Journal of Mathematical Education in Science and Technology*, *43*(6), 703–716. https://doi.org/10.1080/0020739X.2011.646325

Creative approaches to teaching mathematics education with online tools during COVID-19

Sharyn Livy, Tracey Muir, Carol Murphy and Allison Trimble

ABSTRACT

In March 2020, COVID-19 restrictions required our university courses to be offered in online mode only. This quick transition meant that lecturers were suddenly tasked with adapting their classroom-based teaching approaches and materials to the online space. For mathematics teacher educators (MTEs), whose on-campus classes were characterized by participation in interactive activities, the challenge was to devise alternative ways to provide their learners with appropriate mathematical experiences. This article examines how two different MTEs utilized technological tools to adapt their courses online. It looks at the features of these tools and considers how they contributed to students' learning, particularly in terms of the Substitution, Augmentation, Modification and Redefinition (SAMR) model.

1. Introduction

In March 2020, COVID-19 restrictions required our mathematics primary pre-service teacher (PST) education courses to be offered in online mode only. This quick transition meant that lecturers, who may have had little or no experience of teaching online, were suddenly tasked with adapting their classroom-based teaching approaches and materials to the online space. For many staff and students, this proved challenging as new ways of teaching and learning were re-conceptualised. For mathematics teacher educators (MTEs), whose pre-pandemic on-campus classes were characterized by participation in interactive activities, the challenge was to devise alternative ways to provide their learners with appropriate mathematical experiences, including the use of concrete representations of abstract concepts, scaffolding of complex learning content, and engagement in lifelike environments (Means et al., 2014).

This article provides examples of how we, as two MTEs, utilized technological tools to adapt our courses online. The courses were taught across undergraduate and postgraduate primary mathematics teacher pedagogy courses, delivered through both synchronous and asynchronous modes, to tutorial groups ranging from 18–40 PSTs. In particular, we look at the features of these tools and consider how they contributed to students' learning.

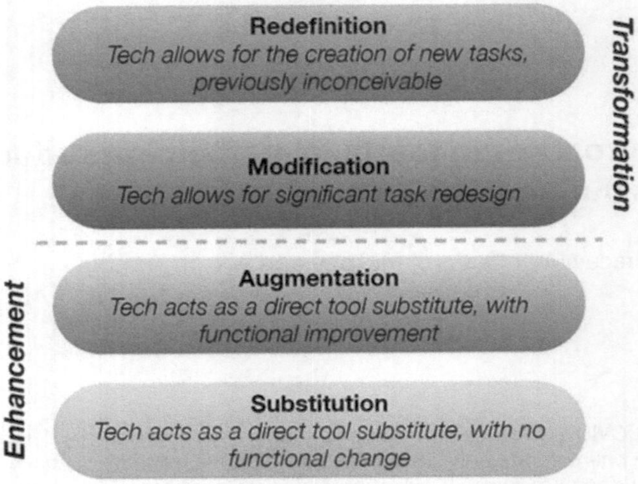

Figure 1. SAMR Model Used for Interpreting our Teaching Approaches.

We found the Substitution, Augmentation, Modification and Redefinition (SAMR) model (Peunedura, 2010), shown in Figure 1, useful for interpreting our teaching approaches.

The SAMR model was designed to classify types of tasks teachers use when integrating technology into their classrooms. Initial integration of tasks suggests enhancement through Substitution and refers to the use of technology to replace a task, with little or no change, such as narrating a PowerPoint presentation. Augmentation occurs when the technology provides additional functions for the task, such as using Kahoot[1], an interactive quiz application. When transforming learning experiences, technology becomes integral to the learning through Modification and Redefinition. Modification occurs when significant changes are made to the learning experience via technology, such as using Google Classrooms or Zoom.[2] Redefinition occurs when the task is adapted and is unimaginable without the use of technology, and cannot be completed without the technology, such as Desmos[3] (Desmos, n.d.).

While we both already made use of technology in our pre-pandemic on-campus tutorials, the COVID-19 restrictions led to searching for, and accessing, technological tools that enabled the implementation of more creative and innovative approaches than were utilized in the past. Next, we provide examples of tasks used when teaching primary (elementary) mathematics education courses to our PSTs. In doing so, we classified our tasks and discuss their features using the SAMR model, beginning with enhancement of learning.

2. Instructional videos and narrated PowerPoint presentations

Prior to the application of COVID-19 restrictions we both provided traditional face-to-face tutorials for PSTs including PowerPoint presentations and demonstration of materials to highlight principles of theory and pedagogy. As a consequence of the restrictions, presentations and demonstrations practices were transferred online. PowerPoint presentations were narrated and made available online. The recorded presentations could be accessed by students anytime, anywhere, with the added benefit of being able to pause and rewind the recording as required (Islam et al., 2020). In addition, bite-sized instructional videos were also produced that enhanced learning experiences by enabling an explicit focus on particular mathematical concepts and skills. For example, a demonstration video was produced showing how to use multi-based arithmetic blocks to model place value concepts. Short 2-3-minute videos of weekly overviews were provided to PSTs by one of the MTEs to maintain a connection with the PSTs and overcome the lack of personal interaction (Brame, 2016; Luo & Kalman, 2018). While the use of PowerPoint presentations and videos were intended as Substitution, arguably the recordings provided Augmentation through functional improvement given the associated affordances such as revisiting and accessing content.

3. Virtual manipulatives

In pre-pandemic delivery modes PSTs extended their knowledge for teaching by using concrete materials including counters, cards, balance beams and dice. Sometimes virtual manipulatives were explored. During COVID-19 we became more reliant on websites and virtual manipulates such as the National Library of Virtual Manipulatives (http://nlvm.usu.edu/en/nav/vlibrary.html) and Illuminations (NCTM, n.d.) (https://www.nctm.org/Classroom-Resources/Illuminations/Interactives/Geometric-Solids/). Such resources were utilized to continue to support and extend PSTs' mathematical learning, and their experiences of teaching with materials.

Educational virtual materials are known to engage primary students and support logical thinking and problem-solving skills (Moyer-Packenham & Westenskow, 2016; Reys et al., 2020), and initially, we used virtual manipulatives to substitute for real materials. For example, virtual pattern blocks[4] (Toy Theatre, https://toytheater.com/pattern-blocks/), shown in Figure 2, were used by the PSTs to construct polygons such as connecting triangles to make, name and classify properties of quadrilaterals. We found that the virtual materials addressed limitation issues associated with real materials, such as having enough materials (pattern blocks) for everyone and being able to take screen shots of a number of different solutions when recording, and documenting responses. In addition, the dynamic features helped PSTs to visualize solutions, experience spatial reasoning skills such as exploring

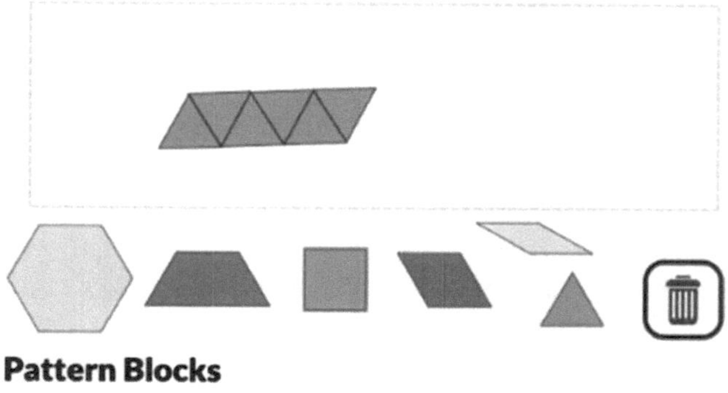

Figure 2. Online Pattern Blocks used to Construct Quadrilaterals.

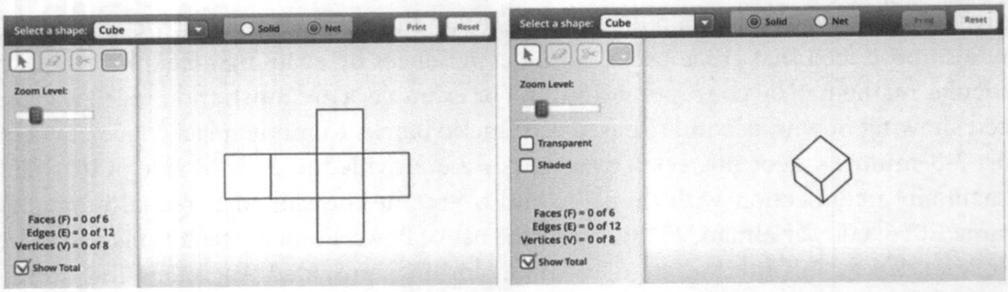

Figure 3. Making Nets and Cubes that can be Flipped and Rotated.

nets for making objects, rotating prisms and cubes to count the number of faces, edges and vertices. Figure 3 Geometric Solids[5] (Illuminations, https://www.nctm.org/Classroom-Resources/Illuminations/Interactives/Geometric-Solids/) an online tool for learning about various geometric solids shows how the polyhedron, cube (on the right) can be rotated to assist with counting faces, edges, and vertices.

Depending upon how virtual manipulatives are used, they can serve as simply a substitute for concrete representations, or the technical features associated with many virtual manipulatives can offer functional improvement and even allow for significant task design. Although the virtual representations were initially introduced as Substitution, these features suggest a level of Augmentation.

4. Cloud based software

Pre-pandemic discussions were managed face-to-face through activities such as Number Talks. Number Talks (Parrish, 2010) evoke rich discussion, but a lot of time in class is required to document everyone's responses, often using a whiteboard. However, when online discussion boards were used for Number Talk activities, students were often reluctant to post, reflecting concerns about the quality of interaction and content generated through the discussion board format (Thomas & Thorpe, 2019).

During the COVID-19 restrictions, solutions were sought to advance engagement with online experiences using interactive cloud-based collaborative web platforms, including Padlet[6] (n.d.). Such platforms allow users to upload and share files to virtual bulletin boards in real time. With regards to Padlet, files are easily accessible, can be anonymous and have the advantage of providing instant visual representation of posts, without having to scroll through discussion threads (Ellis, 2015). Platforms such as Padlet allow PSTs to anonymously post their solutions online, think about their strategy in their own time, and visually see all the class's responses (Fisher, 2017). For example, Figure 4 shows an example of Padlet responses to the Number Talk challenge: 'Do the following in your head $25 + 89$, record your thinking for solving the problem and post to Padlet'.

Another feature and advantage of Padlet (https://padlet.com/) can be seen in Figure 5. In this example, PSTs took a maths shot (photograph) and uploaded it to Padlet, along with a stimulating mathematical question for others to think about. Similar to the Number Talk activity shown in Figure 4, PSTs could view the responses of everyone in the class as well as complete the task at a time that suits them.

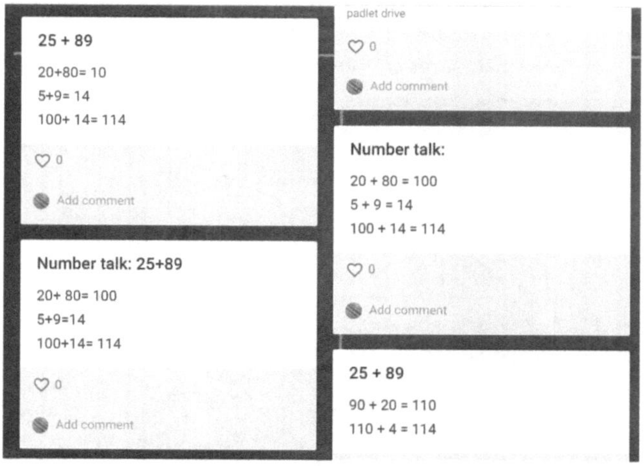

Figure 4. PSTs' Response Posts to Number Talk Activity that can be Shared with each Cohort.

In the online environment, Padlet proved to be an efficient tool for recording multiple responses, in a timely manner, which could be easily read by other students (Lowe & Humphrey, 2018). Padlet was used to upload different files, such as videos, and other affordances of Padlet were capitalized on through following up the activity with a synchronous webinar held via Zoom. In this example, the lecturer used the photos as a stimulus to encourage participants to post additional questions and discuss a range of possible answers (Figure 5).

Hence, although introduced as a Substitution, the combination of Padlet and the online Number Talk activity provided another example of functional improvement, suggesting Augmentation. Further affordances, such as synchronous webinars, provided more significant task design (Modification) as the Zoom platform allowed for remote sharing of questions, ideas, and answers.

5. Collaborative tasks and transformation

Pre-pandemic PSTs in face-to-face tutorials often worked on tasks as a table group and recorded their solutions individually on their own laptops, paper or as a group on one device. During COVID-19 restrictions, group tasks were transferred online, and collaborative approaches were fostered when recording responses to tasks. The PSTs were given a shared link for Google Jamboards (n.d.) that can be set up with multiple pages, providing access to virtual representations alongside synchronous teaching and collaboration (Kimberly, 2020; Sweeney et al., 2021). For example, when learning about geometry, PSTs were given the task of sorting objects, into groups, such as cylinders, prisms and pyramids. Through completing the task using Jamboards and accessing virtual online objects, PSTs were able to post numerous possible solutions (see Figure 6) that did not rely on them having access to the real objects. When synchronous teaching was used in this situation, PSTs completed the activity in Zoom breakout rooms to simulate group work.

Tasks such as Graphing Stories were also introduced using in the programme Desmos, As shown in Figure 7, students first watched a video of a man using a slide, then graphed

Figure 5. Math Shots of Fish Tank and Scoop of Ice-Cream (examples of online Mathematical Problems).

their responses using the software. In one example, PSTs were required to consider two variables (representations) such as the height of a person's waist when climbing up and moving down a slide and the time taken to go up and down the slide. There is also the provision for the MTE to anonymise PSTs' responses, share and discuss the different responses, as well as overlay the graphs to compare responses. Students can then change their graph and finally overlay their answer with the correct response.

This virtual collaborative approach enabled efficient sharing of ideas that is not always achievable in the face-to-face classroom or with large cohorts and allowed for further, transformative teaching, suggesting Modification, as the online environment allowed access to a greater variety of 3-dimensional objects than were available in the face-to-face classroom (or home). For example, in using Jamboard, PSTs could click to the next page to

Figure 6. Using Google Jamboard to Classify Objects in a Venn diagram and Screen Shots of Objects.

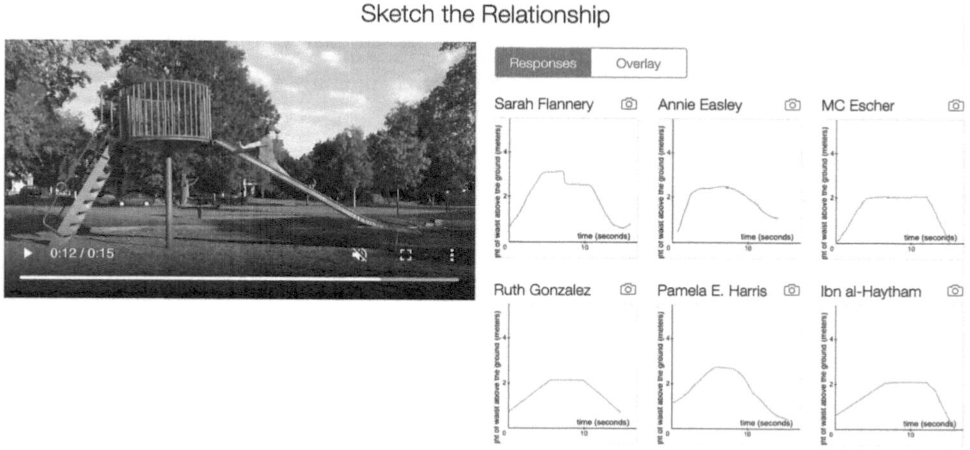

Figure 7. Graphing Stories with Six Responses using Line Graphs.

compare and discuss how different groups responded to the task, sharing their responses and extending their knowledge for teaching (Ball et al., 2008). The Graphing Stories example using Desmos represented the highest level of task transformation (Redefinition) in that the task was both inconceivable and unable to be completed without the use of technology.

6. Takeaways

The rapid shift to online learning and the accompanying challenges that this involved provided the impetus for us as MTEs to examine our teaching approaches and seek out resources and tools that were accessible, creative and effective. The use of the SAMR framework allowed us to critically examine whether or not we were capitalizing on the technology to enhance our online practice. The framework has wide applicability for thinking about online teaching and readers may use it as a reference to consider the transformation or even enhancement of their online teaching.

A takeaway for us as MTEs was that having to substitute physical tools and resources, such as manipulatives, with virtual versions enabled us to combine technology with new pedagogical approaches in ways which actually enhanced, rather than limited, our teaching of primary mathematics. Our experiences with utilizing a range of online tools demonstrate a number of benefits. These include being able to rewind and re-watch demonstration videos, addressing the supply and limited range of physical materials through virtual manipulatives, and accessing more efficient presentation of ideas across larger cohorts, achieved by using tools such as Padlets and Jamboards. We recommend that the creative use of these tools can, and should, be continued post-pandemic.

Notes

1. Kahoot is a game-based learning platform and can be used to create quizzes.
2. Google classrooms and Zoom are both cloud based conferencing service.
3. Desmos is a range of different mathematical activities where students can share their responses.
4. Toy Theatre is a collection of interactive educational games for elementary students.
5. Illuminations mathematics lessons and interactive online tools.
6. Cloud based platform for hosting real time collaboration online.

Disclosure statement

No potential conflict of interest was reported by the authors.

ORCID

Sharyn Livy http://orcid.org/0000-0001-7753-3102
Tracey Muir http://orcid.org/0000-0001-6305-0679

References

Ball, D. L., Thames, M. H., & Phelps, G. (2008). Content knowledge for teaching: What makes it special? *Journal of Teacher Education*, *59*(5), 389–407. https://doi.org/10.1177/0022487108324554

Brame, C. J. (2016). Effective educational videos: Principles and guidelines for maximizing student learning from video content. *CBE—Life Sciences Education*, *15*(4), es6. https://doi.org/10.1187/cbe.16-03-0125

Desmos Graphing Stories. (n.d.). *Desmos Graphing Calculator*. https://teacher.desmos.com/activitybuilder/custom/58797d35d81a612605304b1f?collections=featured-collections%2C5da6476150c0c36a0caf8ffb.

Ellis, D. (2015, October 29–30). *Using Padlet to increase student engagement in lectures*. 14th european conference on e-learning: ECEl2015, 195–198.

Fisher, C. D. (2017). Padlet: An online tool for learner engagement and collaboration. *Academy of Management Learning and Education*, *16*(1), 163–165. https://doi.org/10.5465/amle.2017.0055

Google Jamboard. (n.d). Jamboard. https://jamboard.google.com/.

Islam, M., Kim, D.-A., & Kwon, M. (2020). A comparison of Two forms of instruction: Pre-recorded video lectures vs. Live ZOOM lectures for education in the business management field. *Sustainability*, *12*(19), 8149. https://doi.org/10.3390/su12198149

Kimberly, M. (2020). How to use Google Jamboard for distance learning success. Equip.Learning. Com, April 30, Available from: https://equip.learning.com/google-jamboard-distance-learning-success.

Lowe, T., & Humphrey, O. (2018). A platform for partnership: A technology review of the Padlet sharing platform. *The Journal of Educational Innovation, Partnership and Change*, *4*(1). https://doi.org/10.21100/jeipc.v4i1.706

Luo, S., & Kalman, M. (2018). Using summary videos in online classes for nursing students: A mixed methods study. *Nurse Education Today*, *71*, 211–219. https://doi.org/10.1016/j.nedt.2018.09.032

Means, B., Bakia, M., & Murphy, R. (2014). *Learning online: What research tells us about whether, when and how*. Taylor & Francis Group.

Moyer-Packenham, P. S., & Westenskow, A. (2016). Revisiting the effects and affordances of virtual manipulatives for mathematics learning. In K. Terry, & A. Cheney (Eds.), *Utilizing virtual and personal learning environments for optimal learning* (pp. 186–215). Information Science Reference. Google Jamboard (n.d.). Jamboard. https://jamboard.google.com/.

National Council of Teachers of Mathematics (NCTM). (n.d.). Illuminations: Geometric solids: https://www.nctm.org/Classroom-Resources/Illuminations/Interactives/Geometric-Solids/.

Padlet. (n.d.). It's a beautiful day: Make something beautiful. https://padlet.com/.

Parrish, S. (2010). *Number talks: Whole number computation*. Math Solutions.

Puentedura. (2010). Substitution; Augmentation; Modification and Redefinition. *Hippasus*. http://hippasus.com/rrpweblog, accessed June 2015.

Reys, R., Lindquist, M., Lambdin, D., Smith, N., Rogers, A., Cook, A., Bennett, S., Robson, K., Ewing, B., & West, J. (2020). *Helping children learn mathematics* (3rd Australian ed.). Wiley.

Sweeney, E. M., Beger, A. W., & Reid, L. (2021). Google Jamboard for virtual anatomy education. *The Clinical Teacher*, https://doi.org/10.1111/tct.13389

Thomas, G., & Thorpe, S. (2019). Enhancing the facilitation of online groups in higher education: A review of the literature on face-to-face and online group-facilitation. *Interactive Learning Environments*, *27*(1), 1–10. https://doi.org/10.1080/10494820.2018.1451897

Your online textbook is ready: a shareable, interactive online textbook in response to COVID-19 lockdowns

Peter K. Dunn, Elizabeth A. Brunton and Michael B. Farrar

ABSTRACT

This article discusses the development of an interactive online textbook for undergraduate teaching, in response to moving teaching online due to COVID-19. The development took place initially for a university with less than 5% of students studying online in 2019. The online textbook was incorporated into a large, diverse, multidisciplinary and predominantly first-year undergraduate research design and statistics course. Digital tools were embedded into the textbook to facilitate student engagement in lieu of face-to-face contact, including online crosswords for low-level statistical literacy, drop-and-drag activities, screencasts, exercises, animations, and videos. The textbook was prepared in markdown format. This approach allows for instant updating of material and provides focussed URLs that educators can use to point students directly to specific learning outcomes. Despite the textbook being developed in response to restricted teaching conditions, it has provided a cost-effective resource, targeted at teaching research design and basic statistics concepts, interpretation, and communication. Having returned to face-to-face teaching, the text is now fully integrated into the course. Importantly, the online textbook is freely available for other educators to use, either in its existing form, or by forking (on Github) the textbook sources to create a bespoke online textbook for other courses.

1. Introduction

Since opening in 1996, the University of the Sunshine Coast (USC) has been dedicated to on-campus face-to-face (F2F) teaching. In 2019, only 4.9% of USC course enrolments were online in Semester 1, and 2.2% in Semester 2 (USC Strategic Information and Analysis Unit, personal communication, April 27, 2021). Consequently, the escalation of COVID-19, necessitating moving all USC students to online study during 2020, was highly disruptive for USC.

COVID-19 led to a USC lockdown in March 2020, four weeks into Semester 1 (of a 13 teaching-week semester). USC paused classes during Teaching Week 5 to enable staff to transition teaching materials and approaches to online delivery (Walsh, 2020). For the remainder of the year, classes continued exclusively in online mode. That is, in one week USC moved from almost entirely F2F teaching to exclusively online teaching.

2. Context

USC is a small regional university (18,112 enrolments in 2020; USC, 2021) based in Queensland, Australia, that was unprepared for the sudden shift to online teaching. Anecdotally, few staff had online teaching experience, little appropriate technology was widely available at USC, and most staff were untrained in the available technologies.

In this article, we discuss our response in the context of a predominantly first-year introductory statistics course, *Science Research Methods* (hereafter, *SRM*). This article discusses the production of a free, online textbook (Dunn, 2021), and offers these resources to other instructors to use, or to create their own bespoke, online statistics textbook.

SRM is a large course (over 900 students in 2020) in almost 30 programmes across science (including animal ecology and environmental science), engineering (including mechatronic and civil), allied health (including health science, nutrition and paramedic science), and sport and exercise science (including exercise physiology). The first two weeks discuss research questions and research design; the remainder is a more traditional first-year statistics course (descriptive statistics; confidence intervals and hypothesis testing for common scenarios; simple linear regression), but with a lower mathematical demand than many comparable courses. Statistical software (jamovi (https://www.jamovi.org/) or SPSS) plays a minor role.

Traditionally, students were expected to attend one weekly 2-hour lecture (F2F; lecture recordings available post-lecture) and a weekly 2-hour tutorial (F2F) where students worked through an in-house hardcopy tutorial book. In transitioning *SRM* to online teaching, lectures were replaced with *screencasts* (Dunn et al., 2015), by adding voice-over and annotations to existing lecture slides split into smaller chunks (mostly shorter than 15 minutes). Tutorials were conducted using the video-conferencing software Zoom. This move away from F2F teaching exacerbated a previously-known shortcoming with *SRM*: the lack of a textbook.

Although textbooks had been used previously, none adequately covered the required topics at the required level (including mathematical demand) for the diverse *SRM* cohort; a textbook was last used in 2012. Instead, students were supplied with lecture slides created using beamer and LaTeX, content was delivered using face-to-face lectures, all supplemented with tutorial workshop materials. The emergence of COVID-19 exacerbated the lack of a textbook, as the usual content-delivery mechanism (i.e. the lecture) were not present and many students did not appreciate the screencasts (for example, they were not easy to flick or search through; they required faster internet connections, etc.). The rapid shift to reliance on online learning resources presented an opportunity to address the lack of an appropriate textbook by creating an online textbook.

3. Ideal technological features

Developing an online textbook requires technological tools. Ideally, the technology would offer the author following features and benefits:

- *Output*: Online and hard copy versions should be able to be created from the same source files to maintain consistency between versions and for efficiency.
- *Easy to create*: Early drafts could be easily created from existing resources.
- *Incremental extensions*: The textbook should be able to be easily extended. This allows materials to be updated regularly, rather than having to be completed before being published.
- *Updating*: The textbook should be easily updated to allow for ongoing improvement.
- *Hosting*: Free/cheap web hosting.
- *Interactive*: The textbook should allow the inclusion of interactive features.
- *Tools:* Ideally, the book could be created with open source and free (or inexpensive) tools, since no funding was available.
- *Equations*: Equations would be formatted using LaTeX markup rather than Word equation editor-style editors.

We chose to use **bookdown** (Xie, 2016, 2020) software which uses documents written in a simple markdown markup language, to produce both html and PDF documents (Gruber, 2021). Using **bookdown** allowed for all the above aspirational principles to be adopted.

4. Tools for creating the textbook

Markdown is a simple, plain text-based, human-readable markup language. Markdown documents are easily converted to other formats (including PDF and html) using the free, open-source **pandoc** tool (https://pandoc.org/).

As an example, the markdown text

```
`Markdown` is such a *great* idea!
```

can be converted into other formats using **pandoc**:

- LaTeX: \texttt{Markdown} is such a \emph{great} idea.
- HTML: <code>Markdown<\code> is such a <em>great<\em> idea.
- Word: Markdown is such a *great* idea!

Markdown documents are plain text .md documents (so can be edited in any text editor). They do not offer the full range of features of Word or LaTeX documents, but the simplicity allows for easy conversion between formats. Despite this simplicity, markdown still allows for referencing, citations, images (including images from external sources), tables, cross-referencing, and automatic section numbering.

Using markdown with **pandoc** is convenient using the RStudio environment (RStudio Team, 2020), with the R packages **rmarkdown** (Allaire et al., 2020; Xie et al., 2018; Xie et al., 2020) to manage the conversion of documents between formats with **pandoc**, and **bookdown** (Xie, 2016, p. 2020) to create and publish the book (Figure 1). While RStudio is an environment facilitating use of the R statistical language (R Core Team, 2020), *using or knowing R is* ***not necessary*** *for creating the book.* However, authors who do know R can leverage this to create images and animations using the R language (in .Rmd files).

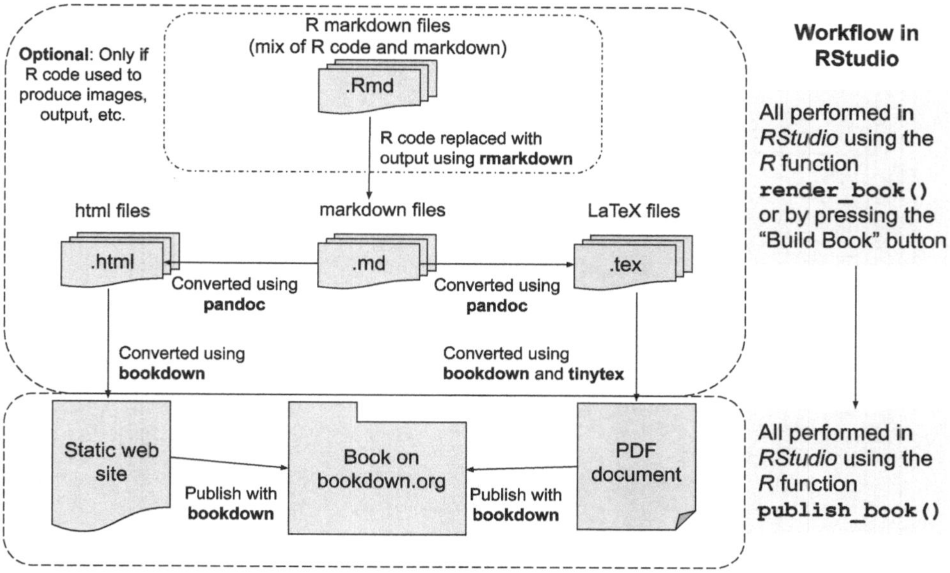

Figure 1. Workflow in RStudio and the underlying processes. If no R code is used, the sources files can be `.md` files rather than `.Rmd` files. Both the online webpages (left side of flowchart) and the PDF version (right side of flowchart) can be hosted on bookdown.org after their creation with **bookdown**.

Using these tools has numerous advantages: (1) all these tools are freely available; (2) all these tools are designed to work together; (3) other necessary tools are automatically installed with these packages; (4) the R statistical language can be used to perform computations (Dunn, 2008), create images and animations if desired; and (5) books created using bookdown may be freely published online at https://bookdown.org/.

5. Creating the textbook in practice

The necessary software (all free and open source) to download and install are:

- **R** (https://cran.r-project.org/).
- **RStudio** (https://www.rstudio.com/products/rstudio/download/), which installs **pandoc** and **rmarkdown** automatically.
- The **bookdown** R package (https://CRAN.R-project.org/package=bookdown).

If PDF output is needed, the **tinytex** R package (Xie, 2019, p. 2020) must be installed also. The required steps are simple and straightforward, and the software is available for most operating systems.

After starting RStudio, create a new 'Project' (*File > New Project > New Directory*), by requesting a new 'Book Project using bookdown'.

This process creates a template for creating a new book (Figure 2), which includes `.Rmd` files ('R markdown files'); each file corresponds to a chapter in the book in the numerical order implied by the file names (Figure 3; left panel). Each chapter contains content in

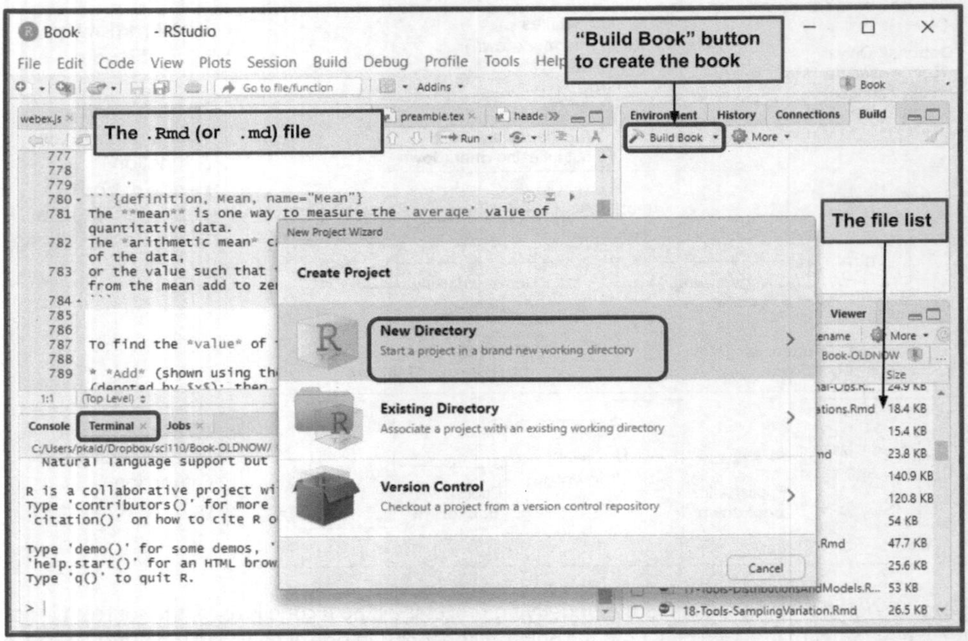

Figure 2. The RStudio window, showing the creation of a new 'Project', the `.Rmd` (R markdown) files, 'Build Book' button to generate the book, the file list, and the Terminal window.

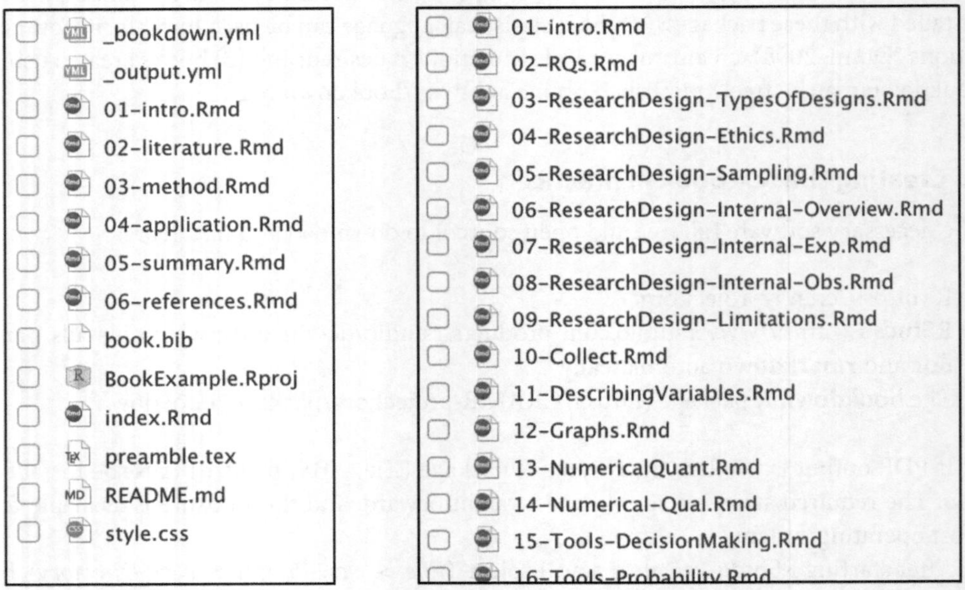

Figure 3. The default files (left panel), and some of the final chapter files for the *SRM* textbook (right panel).

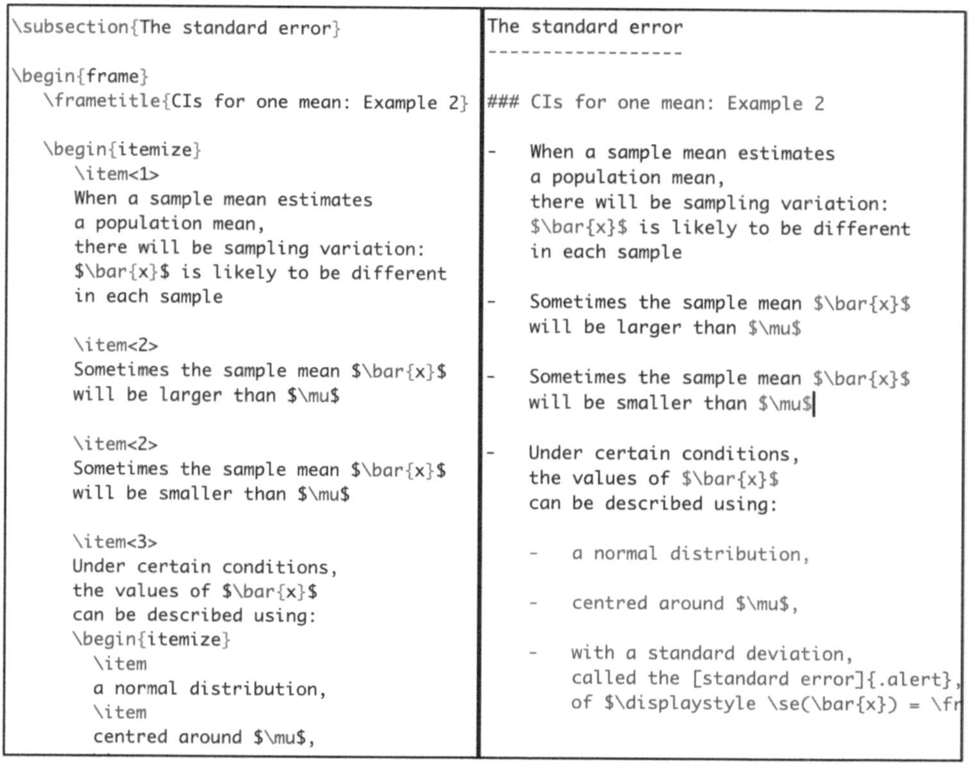

```
\subsection{The standard error}

\begin{frame}
  \frametitle{CIs for one mean: Example 2}

  \begin{itemize}
    \item<1>
    When a sample mean estimates
    a population mean,
    there will be sampling variation:
    $\bar{x}$ is likely to be different
    in each sample

    \item<2>
    Sometimes the sample mean $\bar{x}$
    will be larger than $\mu$

    \item<2>
    Sometimes the sample mean $\bar{x}$
    will be smaller than $\mu$

    \item<3>
    Under certain conditions,
    the values of $\bar{x}$
    can be described using:
    \begin{itemize}
      \item
      a normal distribution,
      \item
      centred around $\mu$,
```

```
The standard error
------------------

### CIs for one mean: Example 2

-   When a sample mean estimates
    a population mean,
    there will be sampling variation:
    $\bar{x}$ is likely to be different
    in each sample

-   Sometimes the sample mean $\bar{x}$
    will be larger than $\mu$

-   Sometimes the sample mean $\bar{x}$
    will be smaller than $\mu$

-   Under certain conditions,
    the values of $\bar{x}$
    can be described using:

    -   a normal distribution,

    -   centred around $\mu$,

    -   with a standard deviation,
        called the [standard error]{.alert},
        of $\displaystyle \se(\bar{x}) = \fr
```

Figure 4. A LaTeX slide converted to markdown after using **pandoc.** (a) An original LaTeX (beamer) file. (b) The same file converted to markdown.

markdown (or Rmarkdown) format. The book can be immediately rendered by clicking one button ('Build Book'; Figure 2). A larger book will have many chapters (Figure 3; right panel), but rendering requires just the press of one button.

Importantly these tools can be used to create *any* book. While mathematical content is easily included, books created using these tools do not have to contain any mathematical or statistical content.

6. Converting existing files

To make the creation of the textbook easier, existing resources slides can be converted to markdown using various online converters. Alternatively, **pandoc** can be used to convert existing Word and LaTeX documents into markdown by typing (in the 'Terminal' window of RStudio; Figure 2):

```
pandoc Document.tex -o Document.md
```

in the folder where `Document.tex` is located (for example). This converts the LaTeX file `Document.tex` into the markdown file `Document.md` (Figure 4), which can be copied into the appropriate textbook markdown files.

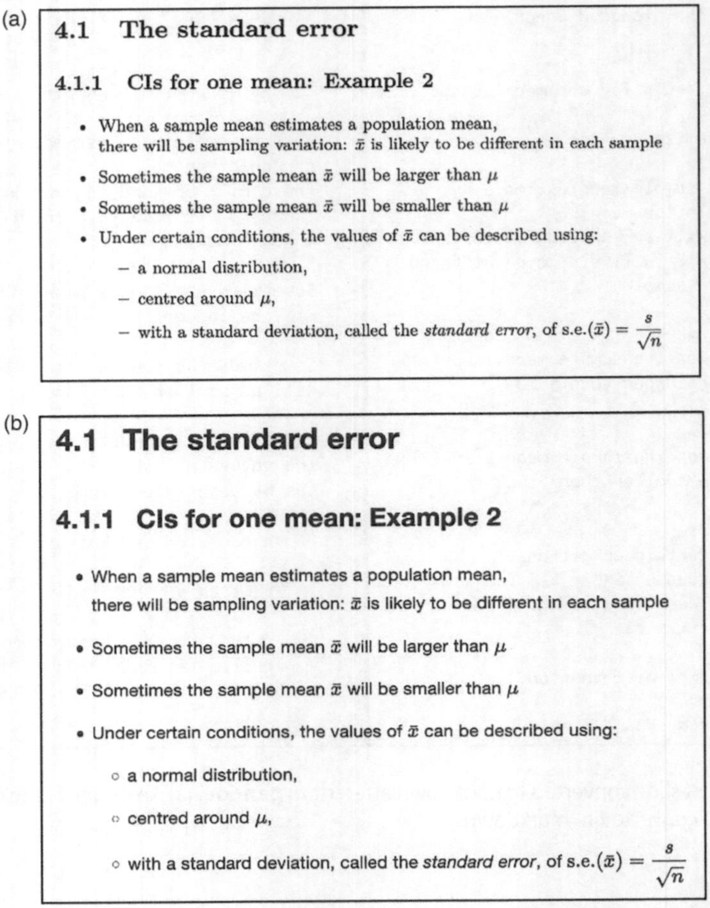

Figure 5. The markdown file in Figure 4 rendered as (a) html and (b) PDF. (a) The html output generated from the markdown source file. (b) The PDF output from the markdown source file.

Usually, only minor manual work is necessary to correct the converted markdown file (for example, to replace the beamer-specific construct `\alert{standard error}` with `*standard error*` in the markdown document, to indicate italics).

The 'Build Book' button (Figure 2) creates either a html book (Figure 5, top panel), a PDF version of the book (Figure 5, bottom panel), or both. Then, to publish the book on `bookdown.org`, use:

```
publish_book()
```

after registering with `bookdown.org`. The html version appears locally in the `_book` sub-directory, as a static website.

Clearly, these instructions simplify the creation of an entire html (or PDF) book. However, even more complex books follow the process described above.

In our context, the first draft of the textbook consisted of basic text, often in dot point format, rendered as html and PDF then gradually developed into a narrative. As time allowed, the book was expanded, streamlined, and features added.

7. Pedagogical features

Pedagogical features of the textbook (some added during 2020) include:

- *A glossary*: A hyperlinked glossary of terms. Glossaries are uncommon in introductory statistics textbooks (Dunn et al., 2017) but useful because of the myriad challenges of learning the language of statistics (Dunn et al., 2016). The glossary entries are cross-linked to the location in the book where the term is mentioned, so provide further context. In this sense, the glossary also acts as proxy index.
- *Explanatory videos and interactivity*: Embedded YouTube, other videos and screen-casts to explain concepts. Online, interactive learning activities, including H5P (https://h5p.com) and online crosswords, can also be included. Other tools could also be used (such as Google forms and Microsoft forms, or from LearningApps.org (https://learningapps.org/)). These artefacts are easy to include. YouTube, for example, provides html code for embedding links into webpages (using the *Share* button below each video), and H5P similarly provides html code for embedding artefacts. This html code can be added verbatim into the markdown files without any further work required. The resource is then added into the created html files when an online version of the book is produced; if a PDF version of the book is created, this html command is simply ignored nothing appears at that location in the book.
- *Real data and studies*: Almost exclusive use of real data and real examples (Neumann et al., 2013). Many of the data files are available in an textbook appendix so that students can work with the data in tutorial or labs. The data and examples come from a variety of disciplines, intentionally drawing on disciplines that take *SRM* (and specifically, ensuring research that studies human and non-human individuals are represented).
- *Notational videos*: Very short videos demonstrating how to *write* and *say* unusual characters with which students struggle (Dunn et al., 2017), such as Greek letters.
- *Exercises and solutions*: Various types of exercises (short answer, multiple choice, computational, conceptual), most with solutions. Some short exercises use drop-down menus for answer options, and give instant feedback, while others have solutions in an textbook appendix. Some solutions were screencasts of a narrator speaking over an animated hand-written solution.
- *Boxes*: Examples and definitions in labelled, formatted boxes (these are built-in **rmarkdown** features) for clarity.
- *Equations*: Equations are easily included using familiar LaTeX constructs. The online version of the textbook includes equations converted using MathML; in the PDF version, the equations are rendered as usual for LaTeX.
- *Branching*: **bookdown** allows for format-specific information (that is, information that may appear in only the html or the PDF version). For instance, the following code chunk will produce the text 'The following short video explains the concept:' in the online version, but no text in the PDF version:

```
`r if (knitr::is_html_output()){
  'The following short video explains the concept:'
  }`
```

This chunk may be followed by the embedding code for a YouTube video, for instance.

- *Images*: Free and free-to-use images (e.g. from Pixabay and Pexels) and icons (Kahlkopf, n.d.) to enhance readability.

7.1. Evaluation

Due to the rapid transition to online-only learning, no plan was made for evaluating the online text in 2020. However, end-of-semester teaching evaluations showed that the online textbook was well received by students. Some aspects of the online textbook that students specifically identified as useful include: interactive questions; worked examples; ease of access; and no cost.

Some examples of students' comments (where the textbook was explicitly mentioned) in the anonymous end-of-semester evaluations include:

- 'The unlimited access to the online textbook and tutorial book was amazing, an absolute lifesaver'
- 'It [the online textbook] also had animations at times, which could help demonstrate an idea more effectively than still images, while being less clunky to open than a short video'.
- 'I would like to highlight that the online textbook [was] remarkable, user-friendly, comprehensive, dependable, flexible, and creative, resources that I would love to see thought-out ALL other subjects in my USC program, regardless of enrolled mode of delivery'.
- 'The idea of an online textbook is great. The book itself was very easy to use + I liked the ability to do the exercises right after the unit'.

Further research is needed to understand how the online textbook experience compares to paper-based textbook experiences for first-year statistics students. Future studies into students' use of different components of the online text across the semester could also inform if student behaviour changes as the course unfolds through research design, to descriptive statistics, to inference.

8. The future: advantages

In 2021, USC returned to F2F teaching, with increased online enrolments (21.8% in Semester 1, a more-than four-fold increase). After returning to F2F teaching, the *SRM* online textbook was retained, updated and formally incorporated into the course as an online and PDF textbook (Dobler, 2015). In 2021 Semester 1, *SRM* (enrolments exceeding 750) uses screencasts in place of lectures and a tutorial workbook, all supported by the textbook, and offers tutorials as either online (Zoom) or on-campus F2F.

The literature reports mixed results using online textbooks (Edgcomb et al., 2015; Gurung, 2017; Henderson et al., 2017). However, in our experience, advantages of using the online textbook for *students* include:

- *SRM* has a specifically dedicated and free textbook.

- Students save money on the textbook and retain access to the textbook when they finish *SRM* and leave USC.
- Students always have access to the latest version of the textbook.
- Students can access the textbook in class on any device.
- The book is easily searchable, unlike videos and screencasts.

Advantages of using the online textbook for *teaching staff* include:

- Any error corrections or enhancements can be made rapidly (without needing to wait for new editions).
- The *SRM* textbook is not constrained to a specific publisher.
- Google Analytics (https://analytics.withgoogle.com/) allows some tracking of textbook usage. This information is restricted to non-identifiable, aggregated information such as the percentage of users in certain locations, the average time spent on various pages by users, and the types of devices being used to access the webpages.
- The textbook is split into numerous small chunks (pages), so URLs can be provided to students (e.g. in emails) to answer questions on specific topics.
- All teaching staff can use the textbook's animations, videos and diagrams in tutorials.
- Teaching staff do not need to lug a hard copy textbook to classes.
- With traditional hard-copy textbooks, only the final version is released. Online textbooks can be released with interim versions; that is, the text can be expanded and enhanced while still being used in the interim.

Looking ahead, one of the statistical software products used in *SRM* (jamovi) will shortly offer an online version (https://cloud.jamovi.org/), opening the possibilities of having jamovi directly embedded in the textbook.

The book has also been available as an open-source resource. We discuss the details of this in the next section; here, we note that open educational resources (OERs) are becoming more popular and are well-received by students and staff (for example, see Hillman et al., 2021).

9. Use our textbook, or make your own

The textbook discussed herein is freely available for anyone to use (Dunn, 2021). In addition, these resources are freely available on Github (https://github.com/PeterKDunn/SRM-Textbook) with a Creative Commons Attribution-NonCommercial-ShareAlike 4.0 International License. This means that corrections to the textbook or suggestions can be made to the textbook. This also means that the textbook source files can be downloaded legally, and forked by any other academic as a starting point for creating their own online textbook, using the free, simple tools discussed. These free tools can be used to create a textbook in *any* discipline and can be freely published online. The process is simple, time-efficient, and provides content and pedagogical tools in a freely accessible, multimedia format. While not essential, mathematical content is easily included so mathematical topics are specifically supported.

These tools offer an opportunity for academics to easily create free, bespoke online textbooks, and overcome technological and logistical barriers to creating online resources.

Disclosure statement

Peter K. Dunn is the author of the online textbook discussed in this article, but stands to gain no financial benefit.

ORCID

Peter K. Dunn http://orcid.org/0000-0003-0299-7169

References

Allaire, J. J., Xie, Y., McPherson, J., Luraschi, J., Ushey, K., Atkins, A., Wickham, H., Cheng, J., Chang, W., & Iannone, R. (2020). *rmarkdown: Dynamic documents for R. R package*. Retrieved May 03, 2021, from https://rmarkdown.rstudio.com.

Dobler, E. (2015). E-textbooks: A personalized learning experience or a digital distraction? *Journal of Adolescent & Adult Literacy*, *58*(6), 482–491. https://doi.org/10.1002/jaal.391

Dunn, P. K. (2008). Building-in quality rather than assessing quality afterwards: A technological solution to ensuring computational accuracy in learning materials. *Journal of Computers in Mathematics and Science Teaching*, *27*(1), 47–63.

Dunn, P. K. (2021). *Scientific research methods: An introduction to quantitative research and statistics in science, engineering and health*. Retrieved 3 May, 2021, from https://bookdown.org/pkaldunn/Book/.

Dunn, P. K., Carey, M. D., Farrar, M. B., Richardson, A. M., & McDonald, C. (2017). Introductory statistics textbooks and the GAISE recommendations. *The American Statistician*, *71*(4), 326–335. https://doi.org/10.1080/00031305.2016.1251972

Dunn, P. K., Carey, M. D., Richardson, A. M., & McDonald, C. (2016). Learning the language of statistics: Challenges and teaching approaches. *Statistical Education Research Journal*, *15*(1), 8–27. https://doi.org/10.52041/serj.v15i1.255

Dunn, P. K., McDonald, C., & Loch, B. (2015). Statscasts: Screencasts for complementing lectures in statistics classes. *International Journal of Mathematical Education in Science and Technology*, *46*(4), 521–532. https://doi.org/10.1080/0020739X.2014.990530

Edgcomb, A., Vahid, F., Lysecky, R., Knoesen, A., Amirtharajah, R., & Dorf, M. L. (2015, June). Student performance improvement using interactive textbooks: A three-university cross-semester analysis. In *the Proceedings of the ASEE Annual Meeting*.

Gruber, J. (2021). *Markdown*. Retrieved May 3, 2021, from https://daringfireball.net/projects/markdown/.

Gurung, R. A. (2017). Predicting learning: Comparing an open educational resource and standard textbooks. *Scholarship of Teaching and Learning in Psychology*, *3*(3), 233. https://doi.org/10.1037/stl0000092

Henderson, M., Selwyn, N., & Aston, R. (2017). What works and why? Student perceptions of 'useful' digital technology in university teaching and learning. *Studies in Higher Education*, *42*(8), 1567–1579. https://doi.org/10.1080/03075079.2015.1007946

Hillman, A. R., Brooks, A. R., Barr, M., & Strycker, J. (2021). Evaluation of open educational resources for an introductory exercise science course. *The International Review of Research in Open and Distributed Learning*, *22*(2), 72–86. https://doi.org/10.19173/irrodl.v22i2.5028

Kahlkopf, A. (n.d.). *iconmonstr*. Retrieved 03 May, 2021, from https://iconmonstr.com/.

Neumann, D. L., Hood, M., & Neumann, M. M. (2013). Using real-life data when teaching statistics: Student perceptions of this strategy in an introductory statistics course. *Statistics Education Research Journal*, *12*(2), 59–70. https://doi.org/10.52041/serj.v12i2.304

R Core Team. (2020). *R: A language and environment for statistical computing*. R Foundation for Statistical Computing. Retrieved May 03, 2021, from https://www.R-project.org/.

RStudio Team. (2020). *RStudio: Integrated development for R. RStudio*. PBC. Retrieved May 03, 2021, from http://www.rstudio.com/.

University of the Sunshine Coast (USC). (n.d.). *Key statistics*. https://www.usc.edu.au/about/vision/key-statistics. Retrieved May 03, 2021, from https://www.usc.edu.au/about/vision/key-statistics.

Walsh, T. (2020, March 20). *Face-to-face teaching paused, but campuses open*. USC News. https://www.usc.edu.au/about/usc-news/news-archive/2020/march/face-to-face-teaching-paused-but-campuses-open.

Xie, Y. (2016). *Bookdown: Authoring books and technical documents with R markdown. Chapman and Hall/CRC. ISBN*, 978–1138700109.

Xie, Y. (2019). Tinytex: A lightweight, cross-platform, and easy-to-maintain LaTeX distribution based on TeX live. *TUGboat*, *40*(1), 30–32.

Xie, Y. (2020). *bookdown: Authoring books and technical documents with R Markdown*. R package version 0.21.

Xie, Y., Allaire, J. J., & Grolemund, G. (2018). *R markdown: The definitive guide*. Chapman and Hall/CRC.

Xie, Y., Dervieux, C., & Riederer, E. (2020). *R markdown cookbook*. Chapman and Hall/CRC. ISBN 9780367563837. URL https://bookdown.org/yihui/rmarkdown-cookbook.

Stop-motion LEGO® animations for learning linear algebra

Emily Cook

ABSTRACT

LEGO® stop-motion animations were developed to engage first-year university students in the topics of matrices, vectors, linear geometry and linear transformations. LEGO® provided a versatile medium through which a wide range of concrete and abstract concepts could be physically demonstrated in both two and three dimensions. It was also familiar to most students, making the mathematics videos accessible and fun. When designed according to key principles around reducing cognitive load in multimedia learning, animations in education have been shown to increase learning compared to static images for concepts involving dynamic systems and processes. Multimodal approaches including video and narration have also been shown to be beneficial to students in terms of demonstrating mathematical problem solving. The videos created in this project were a mix of conceptual explanations and problem-solving examples, where the word problems were physically modelled before being converted to formal mathematical notation and then solved. Some animations were short clips that were integrated into online lecture recordings, others were stand-alone videos designed either as short introductions to topics or examples of common questions. This paper describes the theory underpinning the design of the animations.

1. Context

Learning and teaching abstract topics in linear algebra has always been challenging (Stewart et al., 2019), and became more difficult with the sudden shift to online learning when lecturers lost the ability to gesture, use physical visual aids or draw on large whiteboards. This led to a need for other ways to present abstract concepts, model physical situations and demonstrate three-dimensional problems. The inspiration for the solution described here came from the realization that a matrix can very easily be modelled using a LEGO® plate (see Figure 2 for an example). This led to demonstrating processes such as matrix multiplication and Gaussian elimination using dots and bricks, representing the relevant steps without getting bogged down in numerical calculations. What started as an element of fun in a boring time was well received by students, then expanded to this project to develop stop-motion animations to illustrate concepts in the topics of matrices, vectors, linear geometry and linear transformations.

1.1. Animation theory

Animations have elements of pictorial representation, simulation and movement (Mayer & Moreno, 2003). The benefits of using animations for learning has been widely debated in the literature, with conflicting results reported about their efficacy in helping students construct mental models compared to using static images (Höffler and Leutner (2007), Berney and Bétrancourt (2016)). The diversity of results from individual studies is likely due to the variety of purposes for and characteristics of the animations. Learning benefits depend on several factors such as user control, use of cuing, level of abstraction and field of study. Mathematics is a field of study in which animations have been shown to have a positive benefit to learners (Berney & Bétrancourt, 2016). Animations were found to be particularly helpful when explaining concepts involving processes with time-based changes and work best when they have an expository purpose and do not merely produce extraneous cognitive load by decorating the learning material (Ploetzner & Lowe, 2012). Mayer and Moreno (2003) describe principles to reduce cognitive load in multimodal learning including coherence (reducing extraneous content), signalling (using cues), contiguity (spatially and temporally aligning information) and modality (presenting words as narration rather than as on-screen text to distribute information between visual and auditory channels).

This modality effect, where learning from a video with animation and narration is more effective than learning from a single element alone, has also been applied to problem solving videos in areas of mathematics. The use of problem solving videos, where experts narrate the approach and while visually developing solutions to problems, has been shown to be useful in calculus (Kay and Kletskin (2012), Loch et al. (2014)), statistics (Dunn et al., 2015) and physics (Hall et al., 2017).

Realistic animations were found to be unsuitable to teach mathematical problem-solving skills (Scheiter et al., 2006) so problem-solving can be modelled using hybrid animations (Scheiter et al., 2010). Hybrid animations create a situational model from a word problem, which is then transformed into a mathematical model which can be solved. This applies the principles of 'concreteness fading' (Goldstone & Son, 2005) in which simulation elements start being realistic but by the end of the simulation are idealized, allowing students to better transfer their understanding to other situations. Hybrid animations were used in this work, in some cases to directly model physical situations which then transition to the abstract, and in other cases to represent abstract concepts physically.

1.2. Use of LEGO® in education

LEGO® is a versatile medium designed to promote users to think creatively and reason systematically (LEGO®, 2014). The LEGO® Education program actively creates resources and there is a thriving LEGO® Education Community where teaching and learning resources and ideas are shared (LEGO®, 2021). LEGO® has been used in a variety of ways in STEM education, most frequently with robotics kits which is reported extensively in the literature (for example see the reviews of robotics kits in schools (Benitti, 2012) and in tertiary education (Ruzzenente et al., 2012)). Other creative uses of this medium include in chemistry to model molecules and bonds (Horikoshi, 2020) and LEGO® serious play, which is a facilitated process of communication using LEGO® that was initially developed to support team-building in businesses and has been applied to tertiary education in a variety of ways (James & Nerantzi, 2018).

2. Animation design

The process of creating the animations was incredibly time consuming, with a two-minute video taking, on average, the equivalent of a week's full-time work to complete. Animation production was a by-product of strict lockdown measures which resulted in the author being unable to leave the house for several months in 2020 and again in 2021. This extended period of boredom created a need to entertain both the self and others and provided time to be creative in the design and construction of complicated LEGO® sets. The stages and considerations in the creation of each animation was as follows:

2.1. Identify topic

Animation topics were identified and selected for the following reasons:

(1) Concepts that were difficult to explain through text or diagrams alone. For example, topics such as finding the equation of a plane (Figure 1) or the distance between two lines were difficult to draw clearly on a tablet when teaching online but could better be demonstrated with 3D physical models.
(2) Processes where sequential steps could be demonstrated in a series of images, such as matrix multiplication (Figure 2).
(3) Modelling word problems by physically constructing a situation before translating the model to the abstract mathematics required to solve the problem. An example is a problem involving adding vectors, where a boat is rowed across a flowing river, though a boat was too complicated to animate so was replaced by either a duck or a turtle (Figure 3).
(4) Fun: if a concept can be made more enjoyable using bricks, characters, and terrible puns, then why shouldn't it be? An example is the video on 2D coordinate systems (Figure 4) which included a polar bear, a 'cartesian' bear and an appearance by Descartes himself.

2.2. Identify mode

Animations were either expository (describing a concept), or problem-based (posing and solving a question). They could also be short clips which were embedded in a longer lecture, or stand-alone videos that could be viewed independently. Table 1 shows how the current animations are distributed between these categories.

The short clips were used to reinforce points made elsewhere in lectures with words and diagrams. For example, the 'matrix multiplication' clip was introduced following an example of the process which used numbers. The example involved many steps, a lot of arithmetic and was a long example. In contrast, the animated clip used plates and dots to indicate which rows and columns multiplied together to give which element in the product matrix (Figure 2), reinforcing the steps in the process visually without requiring any

Figure 1. Model demonstrating the normal to a plane.

Figure 2. Still from an animation demonstrating matrix multiplication (the third row of the first matrix dotted with the second column of the second matrix gives the element in the third row and second column of the output matrix).

Figure 3. Still from an animation solving the following problem: A turtle that can swim at 5 m/s tries to cross a river flowing at 3 m/s. What direction should the turtle swim in to reach the point on the bank directly opposite it's starting point?

Figure 4. A polar bear and a 'cartesian' bear, used to represent polar and cartesian co-ordinates respectively.

Table 1. Modes and purposes of existing animations.

	Embedded	Stand-Alone
Expository	Gaussian Elimination Normal to a plane Matrix Multiplication Distance between a point and a line	Linear Transformations (with a sheep) 2D Coordinate systems (with bears) Matrix Multiplication (with a frog)
Problem-Based	Linear Transformations (examples) Linear Transformations (examples)	Adding Vectors (with a duck) Velocity Vectors (with a turtle)

calculation. The clip had no narration with the accompanying audio being some relaxing light jazz.

2.3. Storyboarding

Detailed planning was required to meet the principles of Mayer and Moreno (2003) such as providing the most concise explanations possible. Scripts for the narration were written in advance and plans for the models required to accompany each section were sketched out. Humour was intentionally incorporated, partly as it has been shown to reduce student anxiety in 'dread courses' such as first-year compulsory mathematics units like linear algebra (Kher et al., 1999), and partly as terrible puns are an integral part of the author's teaching style.

2.4. Constructing models

The construction of LEGO® models was fun but also required careful calculations. LEGO® studs are integer values so not all examples that work mathematically could be physically built. Pythagorean triples were used frequently and scales were based on studs (e.g. 4 studs per metre). Single-stud characters were used in problems to make measurements clearer.

2.5. Filming

Photos were captured frame by frame using a phone camera and a free app (Stop Motion Studio). For the problem-solving videos, the solutions were handwritten (Jordan et al., 2012) using a tablet and screen-capturing software (Camtasia). The solutions were later superimposed on a whiteboard in a LEGO® lecture theatre with the narrator personified as a mini-figure® (Figure 5).

2.6. Editing

Animations were edited using Adobe Premiere Rush. Multiple clips were stitched together and supporting structures in the LEGO® hidden by overlaying images. Narration was added, using a conversational tone as advocated by Clark and Mayer (2016). Narration was chosen over written text to allow learners to process words via the audio channel and prevent the visual channel being overloaded. Care was taken to align narration and visuals as students learn better when the corresponding animation and narration are presented simultaneously rather than successively (Mayer & Moreno, 2003). Other features such as arrows and zooming were added as visual clues to highlight key features (De Koning et al., 2009).

2.7. Reviewing

Feedback was sought from subject matter experts on content, pace, clarity and the balance between correct mathematical content and concise delivery. Additions and revisions were made accordingly.

Figure 5. Lecture theatre with lecturer. Calculations were hand-written and superimposed on the whiteboard.

2.8. *Publishing*

Short clips were embedded in lectures and stand-alone animations were published on the learning management system and made publicly available both by sharing in the LEGO® Education community and on YouTube.

3. Conclusion and takeaways

A series of animations have been created to teach concepts in linear algebra. These are freely available on YouTube and can be used directly by educators teaching these topics or serve as ideas for future creative endeavours in mathematics education. While stop-motion animation is a specialist and time-consuming medium, and is thus not being advocated as a method of teaching except by those for which it is a passion, some key points from the creation and implementation of these videos are:

- In topics such as linear algebra it is important to model physical and abstract concepts and problems with concrete representations, before moving towards abstract representation and mathematical notation.
- Creativity can and should be used in mathematics education. Here LEGO® was used as a versatile medium to build physical models and tell stories, but teaching resources which incorporate the teacher's interests are likely to also connect with students and technology provides a wealth of presentation options for content.
- Fun and humour are important in engaging students in mathematical content and in making concepts and terminology memorable. One example of using LEGO® and animal puns is presented here but there many others (for example see Berk (1996)).
- Animations can be used as short clips to break up a longer lecture. These provide both a cognitive break and reinforce topics described in more traditional formats. They do not replace the formal teaching, but reinforce ideas by presenting them in multiple ways.
- Stand-alone animations can be used to introduce topics, where the humour and representations provide a hook to remember new terminology. This removes the element of intimidation when approaching new content for the first time.
- When creating animations, it is important to consider principles such as modality, spatial and temporal contiguity, signalling and coherence.

Future work will include continuing to identify topics where animations can support learning, developing a bank of freely available animations, and investigating which types of animation and features within them most benefit student engagement and learning in terms of conceptual understanding and retention.

Acknowledgements

This work was supported by an Adobe Innovation Grant.

Disclosure statement

No potential conflict of interest was reported by the author.

References

Benitti, F. B. V. (2012). Exploring the educational potential of robotics in schools: A systematic review. *Computers & Education, 58*(3), 978–988. https://doi.org/10.1016/j.compedu.2011.10.006

Berk, R. A. (1996). Student ratings of 10 strategies for using humor in college teaching. *Journal on Excellence in College Teaching, 7*(3), 71–92.

Berney, S., & Bétrancourt, M. (2016). Does animation enhance learning? A meta-analysis. *Computers & Education, 101*, 150–167. https://doi.org/10.1016/j.compedu.2016.06.005

Clark, R. C., & Mayer, R. E. (2016). *E-learning and the science of instruction: Proven guidelines for consumers and designers of multimedia learning.* John Wiley & sons.

De Koning, B. B., Tabbers, H. K., Rikers, R. M., & Paas, F. (2009). Towards a framework for attention cueing in instructional animations: Guidelines for research and design. *Educational Psychology Review, 21*(2), 113–140. https://doi.org/10.1007/s10648-009-9098-7

Dunn, P. K., McDonald, C., & Loch, B. (2015). Statscasts: Screencasts for complementing lectures in statistics classes. *International Journal of Mathematical Education in Science and Technology, 46*(4), 521–532. https://doi.org/10.1080/0020739X.2014.990530

Goldstone, R. L., & Son, J. Y. (2005). The transfer of scientific principles using concrete and idealized simulations. *The Journal of the Learning Sciences, 14*(1), 69–110. https://doi.org/10.1207/s15327809jls1401_4

Hall, B., Loch, B., Rowlands, W., Cook, E., Edwards, T., & Borland, R. (2017). *Will students watch Physcasts if they are not assessed?* Paper presented at the Proceedings of The Australian Conference on Science and Mathematics Education (formerly UniServe Science Conference).

Horikoshi, R. (2020). Teaching chemistry with LEGO® bricks. *Chemistry Teacher International, 1*(ahead-of-print). https://doi.org/10.1515/cti-2020-0017

Höffler, T. N., & Leutner, D. (2007). Instructional animation versus static pictures: A meta-analysis. *Learning and Instruction, 17*(6), 722–738. https://doi.org/10.1016/j.learninstruc.2007.09.013

James, A., & Nerantzi, C. (2018). Guest editors: A potpourri of innovative applications of LEGO® in learning, teaching and development. *International Journal of Management and Applied Research, 5*(4), 153–156. https://doi.org/10.18646/2056.54.18-011

Jordan, C., Loch, B., Lowe, T., Mestel, B., & Wilkins, C. (2012). Do short screencasts improve student learning of mathematics. *MSOR Connections, 12*(1), 11–14. https://doi.org/10.11120/msor.2012.12010011

Kay, R., & Kletskin, I. (2012). Evaluating the use of problem-based video podcasts to teach mathematics in higher education. *Computers & Education, 59*(2), 619–627. https://doi.org/10.1016/j.compedu.2012.03.007

Kher, N., Molstad, S., & Donahue, R. (1999). Using humor in the college classroom to enhance teaching effectiveness in 'dread courses'. *College Student Journal, 33*(3), 400–400.

LEGO®. (2014). *A system for learning.* https://education.lego.com/v3/assets/blt293eea581807678a/bltc6169ff5a6b89a0a/5f86ec84c4805b1a2cea4dfa/lego-education-manifesto.pdf

LEGO®. (2021). *LEGO® education community.* https://community.legoeducation.com/home

Loch, B., Jordan, C. R., Lowe, T. W., & Mestel, B. D. (2014). Do screencasts help to revise prerequisite mathematics? An investigation of student performance and perception. *International Journal of Mathematical Education in Science and Technology, 45*(2), 256–268. https://doi.org/10.1080/0020739X.2013.822581

Mayer, R. E., & Moreno, R. (2003). Nine ways to reduce cognitive load in multimedia learning. *Educational Psychologist, 38*(1), 43–52. https://doi.org/10.1207/S15326985EP3801_6

Ploetzner, R., & Lowe, R. (2012). A systematic characterisation of expository animations. *Computers in Human Behavior, 28*(3), 781–794. https://doi.org/10.1016/j.chb.2011.12.001

Ruzzenente, M., Koo, M., Nielsen, K., Grespan, L., & Fiorini, P. (2012). *A review of robotics kits for tertiary education.* Paper presented at the Proceedings of International Workshop Teaching Robotics Teaching with Robotics: Integrating Robotics in School Curriculum, Riva del Garda (Trento, Italy) April 20, 2012.

Scheiter, K., Gerjets, P., & Catrambone, R. (2006). Making the abstract concrete: Visualizing mathematical solution procedures. *Computers in Human Behavior*, *22*(1), 9–25. https://doi.org/10.1016/j.chb.2005.01.009

Scheiter, K., Gerjets, P., & Schuh, J. (2010). The acquisition of problem-solving skills in mathematics: How animations can aid understanding of structural problem features and solution procedures. *Instructional Science*, *38*(5), 487–502. https://doi.org/10.1007/s11251-009-9114-9

Stewart, S., Andrews-Larson, C., & Zandieh, M. (2019). Linear algebra teaching and learning: Themes from recent research and evolving research priorities. *ZDM*, *51*(7), 1017–1030. https://doi.org/10.1007/s11858-019-01104-1

Crowdsourcing solutions in the online mathematics classroom

Lorelei Koss

ABSTRACT
This article discusses a pedagogical technique for utilizing breakout rooms for small group discussions and synchronous drawing features in Zoom to create crowdsourced solutions to mathematical problems that involve graphing or visualization. Students begin with specific examples and then move quickly to the development of a more general understanding of the concepts. The process was primarily used in synchronous settings, although certain aspects could be utilized in asynchronous settings. The method discussed in this paper incorporates active learning techniques into the online classroom, helps students build community by working in groups, and increases student engagement in the online class meetings.

1. Introduction

There is a wide body of research indicating that active learning techniques help students develop and retain a deep understanding of mathematical concepts (CBMS, 2016; Freeman et al., 2014; Lugosi & Uribe, 2020; Ng et al., 2020; Wieman, 2014) and can help narrow achievement gaps of underrepresented groups in STEM classes (Theobald et al., 2020). The global COVID-19 pandemic forced institutions of higher education to make a rapid shift to online education and required faculty to adapt courses and pedagogies that were traditionally offered in-person to remote teaching formats. Quickly transforming courses that utilized active learning techniques to an online format was challenging. In this paper, we discuss an adaptation of the think-pair-share model (Lyman, 1987) in which students created crowdsourced solutions to mathematical problems that involved graphing or visualization. The technique discussed in this paper incorporates active learning techniques into the online classroom, helps students build community by working in groups, and increases student engagement in class meetings.

The switch to online courses made it tempting to rely heavily on technology as this was the mode of delivery. However, there is a large body of research indicating that outsourcing mathematical processes, called black-boxing, can present risks to students when they become dependent on the tools and do not develop a deep understanding of the underlying processes (Dreyfus, 2006; Jankvist et al., 2016, 2018). As one example, Jankvist and Misfeldt discuss some black-boxing difficulties when using a Computer Algebra System in a differential equations course (Jankvist & Misfeldt, 2015). Their paper relies on mathematics education theory developed by Sfard on three steps students take to master understanding of a new concept: interiorization, condensation, and reification (Sfard, 1991). In the interiorization stage, students become familiar with low level processes and operations. In the condensation stage, students develop the ability to move from thinking about individual steps to thinking about the process as a whole. In the reification stage, the new concept is understood separate from the process that created it.

By crowdsourcing, we mean that student groups work on pieces of a problem that are combined to obtain a larger and more complete result. Unlike the use of crowdsourcing in industry where individual participants may not see how their work influences the final product, the exercises discussed here are designed to help students move from working on their individual problem to a position where they develop an understanding of a larger concept. The goal is to move students through the interiorization and condensation stages of Sfard's schema and minimize black-boxing risks. The examples discussed here were implemented using the software program *Zoom*, and we adopt the naming conventions of that platform (*breakout rooms* for small group discussions, *annotate* for simultaneous drawing), although the techniques can be adapted to other software that allow these two features.

The general procedure is as follows. To begin, the instructor introduces a problem that has been split into smaller pieces and allows students a few moments to think about an approach to investigate their particular problem. Next, students are sent to breakout rooms in small groups to discuss and solve their piece of the problem. These first two steps are to help students become familiar with the techniques and complete the interiorization stage of Sfard's process. Finally, all students return to the main classroom, and one member of the group or all students in the class use the annotate feature to sketch simultaneously on a graph. After students have finished their drawings, the instructor has many options to push the discussion forward. Are there errors in the sketches that need to be discussed? Were some groups given problems that presented unique challenges? Do the combined drawings lead to a general principle that can be conjectured and proved? How would a small tweak to the given problem change the outcome? The resulting sketch and class discussion helps students move through Sfard's condensation stage and in to the reification stage.

In this note, we present how these techniques were used in a Differential Equations course and a Complex Analysis course. Our discussion focuses primarily on planning online activities that promote progression through Sfard's structure and minimize black-boxing risks, but we do not present formal research on the effectiveness of this process, which could be the subject of further study. We conclude with some recommendations on how instructors might plan to use these activities in other courses as well as some brief suggestions on how the instructor can manage these exercises.

2. Examples from differential equations

Differential Equations presents many opportunities for students to draw graphs representing solutions. We present two examples from the most recent semester where students worked synchronously and asynchronously to crowdsource and develop solutions.

Figure 1 shows work that students did to sketch the direction field and solutions of $dy/dx = x - y$. For the first part of the assignment, students worked in groups to sketch the direction field, where each group was assigned a particular quadrant to investigate and were asked to sketch at least 16 mini-tangent lines in that quadrant. This part of the assignment

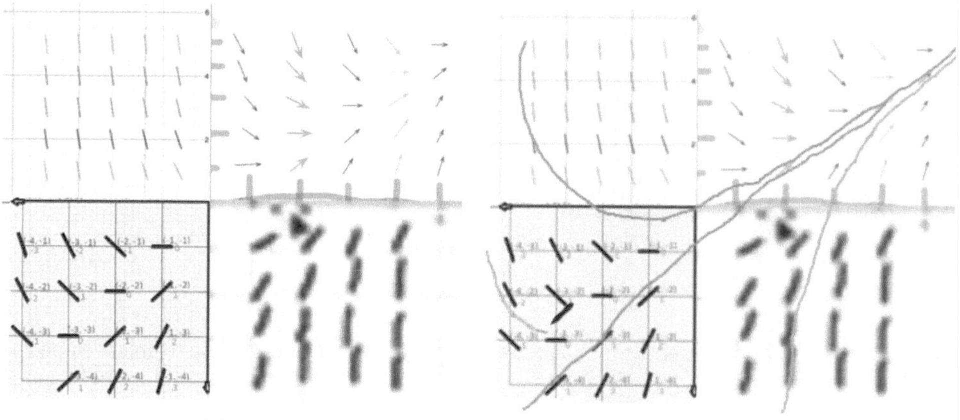

(a) (b)

Figure 1. Constructing a direction field and solutions for $dy/dx = x - y$. (a) Groups worked asynchronously to sketch portions of a direction field and (b) During synchronous class, students sketched solutions simultaneously.

was asynchronous, and groups submitted a short video as homework explaining their solutions. The instructor collated the drawings that the groups submitted and combined four of them to make the entire direction field shown in Figure 1(a). Notice that the group working on Quadrant I drew hash marks on the axes which were not slope lines and thus had to be ignored, and they also put unnecessary arrowheads on the end of the of their mini-tangent lines. Both of these issues lead to productive class discussions. The group working on Quadrant IV sent a blurry figure which made the direction field unsightly but still understandable. During the next synchronous class, students saw Figure 1(a) and were asked to draw possible solution curves. Figure 1(b) shows a screenshot taken while students were using the annotate feature to sketch their group's solutions.

A second example which utilized simultaneous drawing was a discussion of the envelope of the differential equation $(y')^2 - 4xy' + 4y = 0$. As an introduction to the topic, students went to breakout rooms and each room investigated two c values. Figure 2(a) shows the work that students were asked to complete in the breakout rooms. When they returned to the main group, we showed that any line of the form $y = cx - c^2/4$ was a solution to the differential equation for an arbitrary real number c. Students then started sketching the particular lines for their c-values on a common graph. Figure 2(b) is a screenshot of this step in progress. At the conclusion, students could see the envelope $y = x^2$ and discussed how it was also a solution to the differential equation.

Both of the examples shown here were constructed to introduce students to a topic that they had not seen before. It would be very time consuming for one student to do all of these steps on their own. Computers are very good at showing the end result, but not all students obtain a deep understanding when they are just shown the work that the computer produces. Having students work on part of the problem helps them understand how all of the pieces fit together. During the semester, the class did similar exercises to examine families of solutions, phase portraits, and solutions to linear and nonlinear systems.

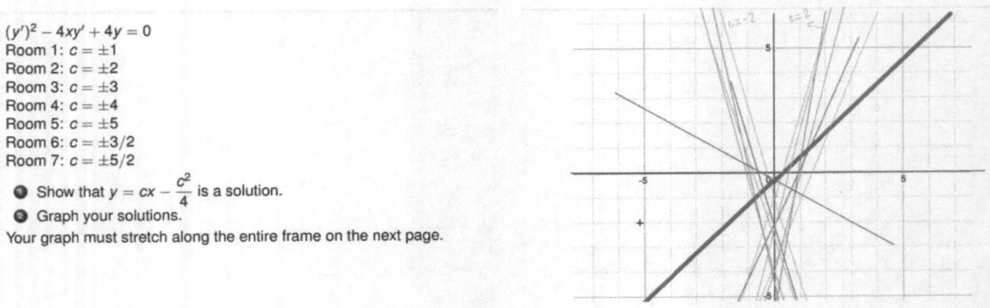

(a) (b)

Figure 2. A screenshot of students simultaneously drawing tangent lines leading to an envelope curve in a differential equations course. (a) Instructions to students and (b) Students drawing simultaneously.

3. Examples from complex analysis

In Complex Analysis, crowdsourcing can assist students in developing a geometric understanding of how functions transform the complex plane. For example, students learned about the complex reciprocal function by investigating how semi-circles were mapped under this function. Students were sent to breakout rooms in groups to investigate a particular semi-circle. After this work was completed, group members sketched the original semi-circle as well as its image under the reciprocal function. Figure 3 shows student work on this problem. After this step was complete, the class developed a conjecture for how an

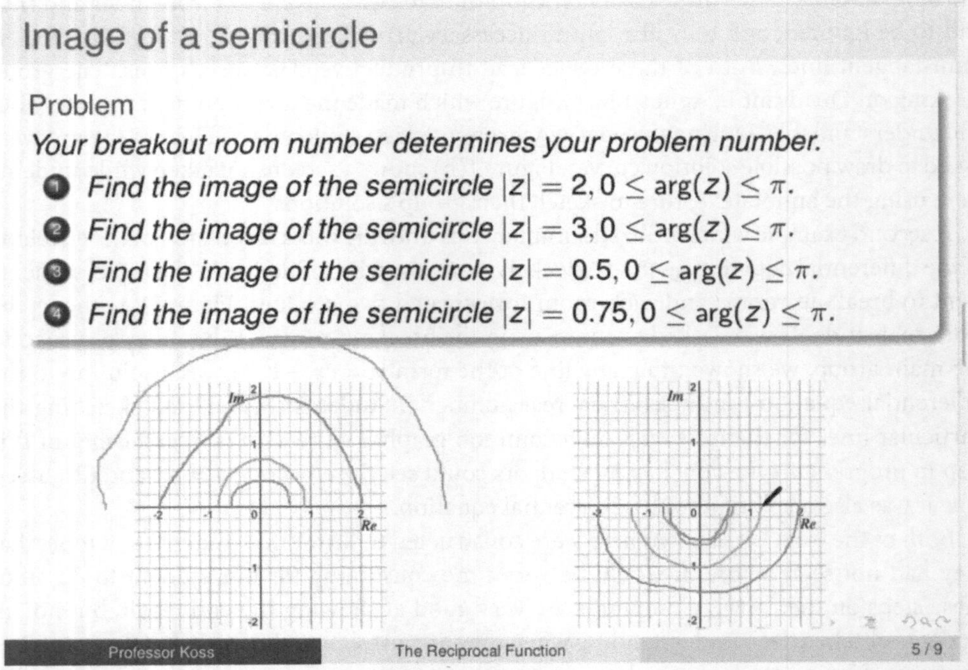

Figure 3. A screenshot of students simultaneously drawing their solutions to how the complex reciprocal function maps semi-circles.

arbitrary circle $|z| = r$ would be mapped and then proved that the conjecture was true. During the same class, students did similar exercises to investigate how the reciprocal function acted on rays starting at the origin, horizontal lines, vertical lines, and half planes.

Over the course of the semester, similar exercises helped students understand the geometric action of complex power functions, the complex exponential function, the complex log function, complex trigonometric functions, linear transformations, linear fractional transformations, and branches of complex roots.

4. Creating similar activities

Although the focus of this paper was on examples from Differential Equations and Complex Analysis, the techniques described could be used in a variety of classes, and we list only a few topics here:

- *Precalculus*: graph transformations, sketching parametric curves, sketching exponential functions with different bases
- *Calculus/Real Analysis:* sketching secant lines converging to the tangent line, sketching derivatives of families such as $f(x) = x^2 + c$, sketching rectangles in Riemann sums, sketching contour diagrams, studying the convergence of a sequence of functions.
- *Discrete Mathematics*: completing complicated truth tables, using truth tables to show logical equivalence, drawing complete directed graphs, drawing regular graphs.

There are a number of ways to select problems suitable for this type of activity. For example, topics which require many tedious steps before students are able to see the larger picture (such as sketching a direction field) are good candidates. Another possibility would be problems for which concrete examples would help build some mathematical intuition before students create a more theoretical argument (such as graph transformations in the complex plane).

The first step is for the instructor to split the tedious or smaller steps into separate activities suitable for groups to investigate independently. The sizes of the groups should depend on the class size, the challenge of the activity itself, and how many steps are required to see the larger idea. Although this paper has used the 'think-pair-share' nomenclature, I have used groups of sizes 2–5 in these activities, and sometimes I have assigned multiple groups the same investigation. For Zoom meetings, groups can contact me if they are stuck and need help getting started. I also drop in unannounced to their groups while they are working. Often, I can tell in a few seconds if they are making reasonable progress, and I leave without speaking to the team. If a group seems confused or off-track, I will interrupt and give them guidance. In this way, the online teaching experience during a synchronous group investigation stage was very similar to what occurs in the classroom.

5. Takeaways

I have done versions of this type of exercise in the physical classroom with mixed success. For example, I have introduced parametric equations in calculus by giving every student a t-value and had them plot the point given by $x(t) = \cos t$, $y(t) = \sin t$ on a large set of axes in the front of the room. The exercise was time consuming; students had to do the computation, walk to the board, find a marker, and physically move around each other to a place where they could draw their point. It was obvious that some students were uncomfortable by their body language. Doing similar exercises on Zoom alleviated many of these problems. For example, students started drawing immediately instead of having to walk to the board, and they didn't have to negotiate the physical space of other classmates. Students could all draw simultaneously, which sped up the process. In addition, completing the drawing over Zoom gave additional anonymity compared to the classroom when everyone was drawing at once. In a suitably equipped classroom, these activities could be performed using shared digital whiteboards or Zoom, even with in-person instruction.

Instructors should think through how they will handle student errors, which should be expected and are shown in two of the examples above. My strategy was to point out that a problem occurred and ask if anyone in the group wanted to talk about what they thought went wrong. For example, when the drawing in Figure 3 was completed, I said that there seemed to be an error in the purple and pink sketches in the image and asked if anyone in those groups saw an error. When errors like these occurred, usually at least one member of the group would offer some comments and try to fix the error. If not, I would move to another group and ask them to explain how they completed their problem. I would then go back to the error and discuss how to complete the task correctly. Finally, it was quick and easy for students to erase and redraw their sketches, which provided an additional check to ensure that students understood how to correct the mistake.

For these types of exercises, using the online simultaneous drawing feature reduces the amount of time between the 'pair' phase and the 'share' phase when students are transcribing their group's work so that the entire class can see it. It was easier for students to correct drawing errors online than it is when they are sketching on the board in the classroom. In this way, using the online simultaneous drawing feature improved the experience for both students and instructor.

Disclosure statement

No potential conflict of interest was reported by the author.

References

CBMS (2016). *Active learning in post-secondary mathematics education*. Retrieved May 29, 2021, from https://www.cbmsweb.org/wp-content/uploads/2016/07/active_learning_statement.pdf

Dreyfus T. (2006). The role of cognitive tools in mathematics education. In R. Biehler, R.W. Scholz, R. Strässer, & B. Winkelmann (Eds.), *Didactics of mathematics as a scientific discipline* (pp. 201–211). Springer Science & Business Media.

Freeman S., Eddy S. L., McDonough M., Smith M. K., Okoroafor N., Jordt H., & Wenderoth M. P. (2014). Active learning increases student performance in science, engineering, and mathematics. *Proceedings of the National Academy of Sciences*, 111(23), 8410–8415. https://doi.org/10.1073/pnas.1319030111

Jankvist U. T., Geraniou E., & Misfeldt M. (2018). The KOM framework's aids and tools competency in relation to digital technologies–a networking of theories perspective. In H.-G. Weigand, A. Clark-Wilson, & A. Donevska (Eds.). (p. 123).

Jankvist U. T., & Misfeldt M. (2015). Cas-induced difficulties in learning mathematics? *For the Learning of Mathematics*, 35(1), 15–20.

Jankvist U. T., Misfeldt M., & Marcussen A. (2016). The didactical contract surrounding CAS when changing teachers in the classroom. *Journal of Research in Mathematics Education*, 5(3), 263–286. https://doi.org/10.17583/redimat.2016.2013

Lugosi E., & Uribe G. (2020). Active learning strategies with positive effects on students' achievements in undergraduate mathematics education. *International Journal of Mathematical Education in Science and Technology*, 1–22. https://doi.org/10.1080/0020739X.2020.1773555

Lyman F. (1987). Think-pair-share: An expanding teaching technique. Maa-Cie Cooperative News, 1(1), , 1–2.

Ng O. L., Ting F., Lam W. H., & Liu M. (2020). Active learning in undergraduate mathematics tutorials via cooperative problem-based learning and peer assessment with interactive online whiteboards. *The Asia-Pacific Education Researcher*, 29(3), 285–294. https://doi.org/10.1007/s40299-019-00481-1

Sfard A. (1991). On the dual nature of mathematical conceptions: Reflections on processes and objects as different sides of the same coin. *Educational Studies in Mathematics*, 22(1), 1–36. https://doi.org/10.1007/BF00302715

Theobald E. J., Hill M. J., Tran E., Agrawal S., Arroyo E. N., Behling S., Chambwe N., Cintrón D. L., Cooper J. D., Dunster G., Grummer J. A., Hennessey K., Hsiao J., Iranon N., Jones II L., Jordt H., Keller M., Lacey M. E., Littlefield C. E., ...Freeman S. (2020). Active learning narrows achievement gaps for underrepresented students in undergraduate science, technology, engineering, and math. *Proceedings of the National Academy of Sciences*, 117(12), 6476–6483. https://doi.org/10.1073/pnas.1916903117

Wieman C. E. (2014). Large-scale comparison of science teaching methods sends clear message. *Proceedings of the National Academy of Sciences*, 111(23), 8319–8320. https://doi.org/10.1073/pnas.1407304111

Leveraging the power of Google Apps to support active learning in a synchronous online environment

Patrick Sullivan

ABSTRACT
A significant challenge of pandemic teaching was creating active small group learning experiences in a synchronous online learning environment. Efforts to address this challenge led to an instructional process, using multiple Google Apps, that was not only effective in an online setting, but has the potential to improve seated small group learning, as well. In this article, a complete implementation of this process will be shared describing how multiple Google Apps (i.e. Forms, Sheets, Slides, Jamboard) were utilized to gather information about the current level of students' understanding, engage groups in a learning activity to advance this understanding, provide immediate feedback as students worked through the activity, and equip the instructor and groups with tools to interact without disrupting small group dynamics.

1. Introduction

Effective small group active learning experiences were a staple of this instructor's seated classroom environment. The instructional process guiding these experiences consisted of four elements. These elements included: the instructor providing groups with a problem set to work, listening to each group's interactions, answering student questions, and asking assessing and advancing questions to move each group's thinking forward. Shifting to a synchronous online teaching environment mid-semester led to the challenge of finding and implementing technology tools that would support a virtual instructional process with similar elements as the seated classroom environment.

Michaelsen (1998) argued that effective use of group learning in an online environment requires three features. These features include accountability for the individual and group learning, assignments that promote both academic learning and group development, and frequent and timely feedback. Incorporating these features into a synchronous online statistics and probability content course for future middle school and high school teachers required several Google Apps. Specific applications used were Google Forms, Sheets, Slides, and Jamboard. A case study (Creswell, 1998) was conducted to identify practices that supported a successful implementation. Multiple sources of data were collected

The local health department wants to investigate whether there is an association between eating at a fast-food restaurant and gender. They conducted a survey of 100 randomly selected people and ask each person the following question, "Do you eat at a fast-food restaurant at least once a week?"

Gender	Yes	No	Total
Male	40	20	60
Female	14	26	40
Total	54	46	100

1. Among all people surveyed, what percentage said they eat at a fast-food restaurant at least once a week?
2. Among males surveyed, what percentage said they eat at a fast-food restaurant at least once a week?
3. Does there appear to be an association between gender and eating at a fast-food restaurant at least once a week?

Figure 1. Google Form questions.

through field notes, student surveys, and assignments. In this article, a complete implementation of one iteration of the instructional process describing how each application was best utilized will be shared. Also, specific tips that support a successful usage of each application in the implementation will be shared.

2. Google Forms

Google Forms was utilized to assess the nature of students' current understanding. As a mathematics teacher educator, a responsibility is to model effective mathematics teaching practices for preservice teachers (National Council of Teachers of Mathematics, 2014). One of those teaching practices is to 'elicit and use evidence of student thinking' to advance student learning. Google Forms was utilized as a tool for creating formative assessments to evaluate the nature of students' current understanding. Google Forms allowed for both closed questions (e.g. multiple-choice) and free-response (e.g. short-answer or paragraph) questions to be asked. An example of a set of questions that were part of a formative assessment is shown in Figure 1.

3. Google Sheets

After students completed the Google Form questions their responses were analysed by the instructor using two features, summary and spreadsheet. This analysis served as the foundation for the learning activity that was created later using Google Slides. The summary feature was helpful because responses to multiple-choice questions are shown in a pie chart. Questions that reflected a wide variation in answers were identified and added to the group learning activity.

The spreadsheet feature allows for the creation of a Google Sheet that is a compilation of all students' responses. This option enabled the instructor to identify themes in the class' reasoning that needed to be explored further. For example, student responses to the second

Gender	Yes	No	Total
Male	40	20	60
Female	14	26	40
Total	54	46	100

Among the males surveyed, what percentage said they eat at a fast- food restaurant at least once a week?

Several students commented that the answer was 0.40 (40%). What is the error in their reasoning? Explain.

Figure 2. Short-answer question.

question in Figure 1, 'Among the males surveyed, what percentage said they eat at a fast-food restaurant at least once a week' revealed a lack of consensus (i.e. 40 vs. 67). Student responses to open-ended questions were also analysed using the Google Sheet feature so that misconceptions and gaps in understanding could also be identified. Two themes identified were that students often confused marginal and conditional probabilities and they mistakenly believed that the independence of two variables allowed for casual inferences to be made.

Though not an aspect of the activity shared in this article, the instructor also had the option of creating a link to share the entire spreadsheet of student responses with students. Columns with student identifiers can be hidden so that only responses to questions appear. (Note: As one of the Google Form questions students are asked to create a four-digit code for the semester that only they and the instructor know. Although the results in the spreadsheet are anonymous individual students can identify their work using the four-digit code.) Once anonymity was established the instructor created reflective assignments that incorporate the class' responses. This type of activity is shared briefly in Figure 4.

4. Google Slides

After the instructor finished analysing student responses, Google Slides was utilized to create the group learning activity. It was easy for the instructor to add screenshots and text boxes for students to respond to questions (Note: Use page setup to create a custom setting of 8.5 by 11 so students have more room for their responses.) Two different types of group learning activities were created by the instructor. The first type, whose implementation is shared in this article, involved incorporating responses to multiple-choice, or short-answer questions, in which there was a lack of class consensus, and responses to free-response questions that revealed gaps in understanding or important misconceptions. An example of each type of question that was incorporated in the learning activity is shown in Figures 2 and 3.

The second type of activity, not discussed in this article, challenged students to put themselves in the role of a teacher reflecting on the nature and quality of student responses. An example of this type of activity is shown in Figure 4.

Does there appear to be an association between gender and eating at a fast-food restaurant at least once a week?

Gender	Yes	No	Total
Male	40	20	60
Female	14	26	40
Total	54	46	100

Evaluate each students' thinking. Provide feedback that would support growth in their understanding.

"When looking at the numbers, we can make and cause and effect statement since there are more men that ate out than women. Men are 40% while women are at 14%. Which is a big difference between the two."

"There does appear to be an association between the males and eating at a fast-food restaurant at least once a week. The table shows us that 74% of the people who ate out once a week were men and only 24% were women. This tells us that men are almost 3 times more likely to eat out once a week compared to their opposite gender."

"Yes, males are twice as more likely to eat at a fast-food restaurant at least once a week than females. The percentage of males were also above 54% and if there was no association between gender and eating fast-food the percentage of males and females would both be 54%."

Figure 3. Free-response question.

	Eye Dominance			
Hand Dominance		Right	Left	**Total**
	Right	54	26	**80**
	Left	6	14	**20**
	Total	**60**	**40**	**100**

1. As a group review the responses to the Hand Dominance Activity (link in Blackboard). Evaluate the correctness of the work and the quality of the written responses.
2. Address the following questions: A) What misconceptions do some students seem to have? B) How would written responses be "tweaked" to be exemplary? C) Identify two or more (with four-digit code) responses that your group considered to be exemplary.

Figure 4. Reflecting on student responses.

Creating a shared Google Folder for the class made it easy for students to access the Google Slides activity. There are two reasons why this is important. After students are assigned to breakout rooms in the synchronous online environment (e.g. Zoom), they need a way to quickly access a shared copy of the Google Slide activity. The instructor created a link to the Google Folder so that 'anyone on the internet with this link can edit'. Inside this folder copies of the Google Slide activity corresponding with the number of desired groups were created and placed in the Google folder. The link to the Google folder can then

Does there appear to be an association between gender and eating at a fast-food restaurant at least once a week?

Gender	Yes	No	Total
Male	40	20	60
Female	14	26	40
Total	54	46	100

Evaluate each students' thinking. Provide feedback that would support growth in their understanding.

"When looking at the numbers, we can make and cause and effect statement since there are more men that ate out than women. Men are 40% while women are at 14%. Which is a big difference between the two."

> ***We can say that there is an association between genders and eating out once a week because male and female did not have the same probability of eating out or not. Out of the 60 men surveyed 40 answered yes, P(y/m))=67%, while out of the 40 women surveyed 14 answered yes, P(y/w)=35%. We can see that these probabilities do not have close to the same value so we can say there is no association between the two.***

"There does appear to be an association between the males and eating at a fast-food restaurant at least once a week. The table shows us that 74% of the people who ate out once a week were men and only 24% were women. This tells us that men are almost 3 times more likely to eat out once a week compared to their opposite gender."

> ***We can see that 54% of all surveyed said they eat out at least once a week. But when we break it down and look at individual genders, we see that males eat out twice as much as females (m = 67%, f = 35%). Neither of these values fit around the total 54% so we can see an association with genders and eating out.***

"Yes, males are twice as more likely to eat at a fast-food restaurant at least once a week than females. The percentage of males were also above 54% and if there was no association between gender and eating fast-food the percentage of males and females would both be 54%."

> ***There is an association. We see the total number of people who said yes, they eat out is 54%. For us to see no association to gender and eating out we would assume that males and females would be the same. When taking the probability of getting a yes given male we get 67%. When taking the probability of getting yes given female we get 35%. Males are twice as likely to eat out.***

Figure 5. Group response to a free-response question.

be shared with the class in the chat feature of the synchronous online environment. Once students have access to the folder, they can choose the activity number that corresponds with their breakout room number (e.g. G3_Fast Food is to be completed by breakout room 3). A nice feature of this is that each student in the group has a live version of the activity on their screen. An example of a group's written response to a free-response question (see Figure 3) is shown in Figure 5.

A second reason why it is important that each group has a separate Google Slide activity is that it enabled the instructor to provide immediate feedback to each group without disrupting student-to-student discussions within the group. A major criticism of students in synchronous online settings is the inability to receive timely feedback (Mogavi et al., 2021). One way to provide timely feedback in group collaborations in a synchronous online

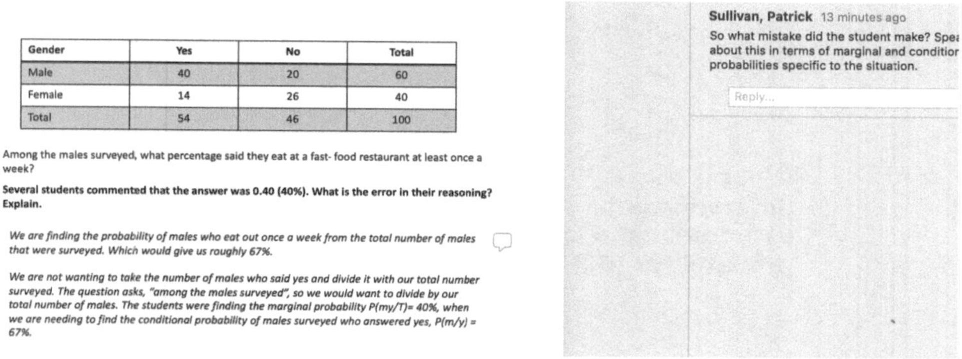

Gender	Yes	No	Total
Male	40	20	60
Female	14	26	40
Total	54	46	100

Among the males surveyed, what percentage said they eat at a fast- food restaurant at least once a week?

Several students commented that the answer was 0.40 (40%). What is the error in their reasoning? Explain.

We are finding the probability of males who eat out once a week from the total number of males that were surveyed. Which would give us roughly 67%.

We are not wanting to take the number of males who said yes and divide it with our total number surveyed. The question asks, "among the males surveyed", so we would want to divide by our total number of males. The students were finding the marginal probability P(my/T)= 40%, when we are needing to find the conditional probability of males surveyed who answered yes, P(m/y) = 67%.

Figure 6. Inserting comments to improve the group's response.

platform (e.g. Zoom) is to enter the group's breakout room. An observation, however, made by the instructor in the early implementations of this instructional process was that student–student collaborations were disrupted when the instructor entered the room. Creating the Google folder with a Google Slide activity for each group afforded the opportunity to see each group's work without seeing them. The instructor would open a tab for each group in his browser. Clicking between the tabs, he could quickly and frequently, using the comments feature of Google Slides, move between each group, observing progress and providing immediate feedback. This also provided each group with a written record of the teacher-group interaction that could be referenced later. This also made it easy for the class to do a gallery walk through each group's work if desired by the instructor. An example of the instructor inserting comments is shown in Figure 6. The group's initial response lacked specificity as to the types of probabilities, so the instructor added a comment which led to the group revising their response (see Figure 6).

5. Google Jamboard

Google Jamboard was added to the instructional process after receiving feedback from students. After an earlier implementation, students indicated that they preferred a large group discussion following small group learning activities to summarize the activity and address any lingering questions. Jamboard was incorporated to provide students with a platform to pose questions that would serve as the foundation for these large group discussions. Using the 'sticky note' feature groups could pose questions to be discussed further (see Figure 7). Another way that Jamboard was utilized by the instructor was to gather feedback from students. The instructor would use Jamboard after a lesson to learn about what resonated with students and burning questions they still had. Jamboard responses allowed the instructor, as well as students, to identify themes that served as the foundation for future class discussions.

6. Evaluation

Overall, students reacted positively to this instructional process. The impact is best exemplified by the student email sent to the instructor.

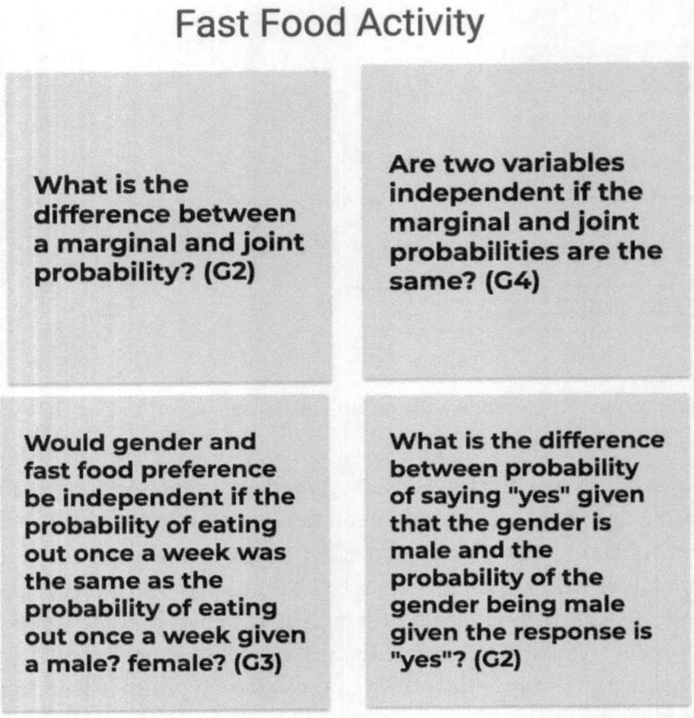

Figure 7. Jamboard questions.

> I wanted to say thank you for class today. It was the first time in quite a long time I consciously noticed myself taking an active role in my learning rather than participating more passively. It was so refreshing to be so mentally engaged and feel as though I was actually holding myself to challenging thought.

At the completion of the learning activity, students were asked to respond to the following question, 'As you reflect upon this learning experience, what do you believe were the most helpful aspects of this experience?'. Several themes emerged from an analysis of student responses to this question. Each theme as well as a student comment that best supported the theme will be shared. The first theme that was identified was that there was value in students being able to seamlessly see the work, in real time, that their peers were adding to the slides and that they could interact with each other while doing so.

> I think it was most helpful that everyone could be viewing the slides, making changes, and leaving comments. This allowed the group to flow more smoothly without having to share screens. It also allowed for feedback to happen while students worked, so they could go back and make adjustments.

The second theme is that they appreciated that the instructor had provided them an opportunity to improve their initial work, 'I think the most helpful thing was you going back through all our slides and adding questions to each slide. I liked how we all had our individual feedback and were able to think even more about the problems.'

The third theme is that students seemed to value collaborating with their peers to deepen their understanding of the content relying on the instructor as needed. In other words, the sources of authority lied within group members and shifted to the instructor only when called upon:

> Looking back at certain questions again in small groups was very beneficial because I was able to reflect on my initial answer and my initial understanding of the question. Seeing what the most common answers were was a great way to start a discussion of each question within my group. Then as we all talked through it, we gained a stronger understanding but when we were still unsure, so we were able to invite you into our group and ask you a question to clarify what we were thinking. Then we were able to come to a conclusion.

7. Implications for post-pandemic teaching

Efforts to implement an active learning instructional process in a synchronous online setting like those that had previously occurred in a seated classroom led to several enhancements. These enhancements involve the nature of feedback and how student questions and thinking is recorded. Compared to the nature of feedback in the traditional seated classroom, this instructional process seemed to have several advantages. First, this process seemed to be more targeted to group needs and the feedback more frequent because each group's written work was more accessible to the instructor. Instead of walking around the room, disrupting student–student collaborations, the instructor could view from their computer the work of each group.

Another option was for the instructor to use the comments feature of Google Slides to add feedback without entering the group's workspace. An additional advantage of using the comments feature is that there is a written record of the feedback to groups. In the instructor's seated classroom this has traditionally not been the case. Typically at the end of seated small group discussions, the instructor would ask students to share if they had any additional questions. This created several challenges. Often these questions were not fully elaborated by students, or students were not willing to publicly share, or the instructor was caught off guard by the nature of the question. Incorporating Jamboard alleviated these challenges. Before posting a question, the group must fully elaborate in written form the nature of the question they desired to ask. Also, submitting these questions in advance of the large group discussion the instructor has time to mentally prepare a response. This is also a more equitable practice because by anonymously posting questions, each student could post questions and no student has to worry about the feelings associated with making a question public.

Another enhancement is that both the instructor and groups had a record of their reasoning that could be accessed to support future discussions. In the seated classroom instructional process, a written record of the important learning that occurred was often lacking within these small group learning activities. As a result, when class ended the fruit of those discussions was sometimes lost and unable to be recovered for future classes. Incorporating this process addressed this challenge because the group's responses, the instructor's questions, as well as additional group questions were recorded.

8. Conclusion

A significant challenge of pandemic teaching was creating active small group learning experiences in a synchronous online learning environment that resembled what was happening in a seated classroom. Addressing this challenge led to the development of an instructional process using multiple Google Apps that supported positive small group active learning experiences in an online environment. Analysis of multiple implementations of this instructional process led to enhancements that support better group collaborations, instructor-group feedback interactions, and recording of student reasoning and questioning. These enhancements, along with how the applications were used in an online setting, are easily transferable to the seated classroom environment.

Disclosure statement

No potential conflict of interest was reported by the author.

ORCID

Patrick Sullivan http://orcid.org/0000-0003-3680-2526

References

Creswell, J. W. (1998). *Qualitative inquiry and research design: Choosing among five traditions*. SAGE.
Michaelsen, L. K. (1998). *Professional and organizational development network essay series teaching excellence: Towards the best in the academy* (Vol. 9). POD Network.
Mogavi, R. H., Zhao, Y., Haq, E. U., Hui, P., & Ma, X. (2021). Student barriers to active learning in synchronous online classes: Characterization, reflections, and suggestions. *Reflections, and Suggestions*. https://doi.org/10.1145/3430895.3460126
National Council of Teachers of Mathematics. (2014). *Principles to actions: Ensuring mathematics successes for all.*

OPEN ACCESS

Flexible, student-centred remote learning for programming skills development

Peter Rowlett and Alexander S. Corner

ABSTRACT

During the COVID-19 pandemic, the teaching of programming for undergraduate mathematicians was moved online. This was delivered asynchronously, with students working through notes and exercises and asking for help from staff via online messages as needed. Staff delivery time was redirected from content delivery into a formal system of formative assessment, which replaced informal discussion and feedback during in-person classes. Formative tasks were submitted and feedback was provided via GitHub Classroom. Students were broadly positive about the formative feedback system and mixed about the need for live delivery. Formal formative feedback highlighted that students may hold incorrect views about the accuracy of task completion, making formal formative submission an effective use of staff delivery time.

1. Introduction

When teaching at UK universities moved suddenly online in March 2020 during the COVID-19 pandemic, there was a need to quickly prepare remote teaching. A little more planning was possible for the 2020/21 academic year running from September 2020 to June 2021.

With teaching that moved online that autumn, there were two forms of delivery dominant in the discussion about how to do this:

- pre-recorded videos followed by live online tutorials for students to get support while completing exercises;
- live online classes offering a mixture of lecturer delivery and student activity.

This paper discusses how this context affected teaching for a second-year optional module in a UK undergraduate mathematics degree, 'Programming with Mathematical Applications', including a decision to offer neither of these delivery modes and instead use module staff time for an enhanced program of formative assessment.

Taras (2005) has formative assessment being an evidence-based judgement accompanied by feedback indicating the gap between a piece of work and a particular standard, and an indication of how the work can be improved to reach that standard. Schoenfeld (2015) has formative assessments as 'examinations or performance opportunities the primary purpose of which is to provide students and teachers with feedback about the student's current state, while there are still opportunities for student improvement'.

By presenting the way this was designed and some reflections on its delivery, we hope to contribute to a discussion about engaging flexibly with remote delivery modes and choosing the right approach for the content being delivered.

2. Module design

In their first year, the students studied 'Mathematical Technology', a compulsory module covering various software packages including a brief introductory section on programming. Some students have other prior programming experience, either from school or from personal interests.

The module has a focus on mathematical aspects of programming and provides experience with multiple languages, covering fundamental programming concepts, graphical interfaces, databases and presenting mathematics on the web (Rowlett, 2020). Teaching is delivered via structured notes giving explanations, examples and exercises. For students who are more experienced with programming, a series of challenge exercises are provided to encourage practice. Pre-COVID-19, class time was arranged with a small amount of delivery highlighting key concepts, but most of the time in class being student-centred, with students working on programming tasks and seeking help and feedback from module staff as required.

The motivations for this individualized approach are:

- with diverse prior knowledge and experience of programming, it is not practical to deliver whole-class content appropriate for all students at once; and,
- programming is a practical skill best learned by doing.

The idea is that students use the available learning resources in a way they feel will best aid their development. In response to the notes and exercises, students are expected to write code and this is intended as the performance opportunity which will prompt a feedback interaction. First, the computer will act on the code the students have written and the response may not be as expected. Second, interaction with module staff might occur, either because the student seeks help or because staff casually enquire about performance. This feedback interaction provides information specifically relating to the task, meaning it is formative feedback with an instructional purpose (Hattie Timperley, 2007). In response to this feedback interaction, students are encouraged to reflect on their coding practice and adjust their learning appropriately.

Students are not, in this module, encouraged to work collaboratively in a formal sense. For example, working with others can lead to a situation where some (perhaps more experienced) students will take over responsibility for some aspects of the work, meaning that others do not benefit from development in these areas. This may be welcome in some areas since learning to work to one's strengths is a useful skill, but in the case of fundamental programming skills, it was felt students would benefit from a more thorough grounding.

Students do work collaboratively in other modules, and of course, informal interaction between peers does take place.

The module is assessed through coursework and an in-depth project. The coursework uses well-defined tasks to test taught content. The project takes place over the last six of 24 teaching weeks and allows students to demonstrate a broader range of skills around designing and managing an individual programming project. During the project, there is no direct teaching; rather the scheduled sessions are an opportunity for students to work independently, seeking help and feedback from module staff as required.

3. COVID-19 disruption in 2019/20

The COVID-19 pandemic and associated restrictions took effect starting in March 2020. A decision was made in the evening of 16th March 2020 that teaching at Sheffield Hallam University would be moved to remote delivery effective the following day, ahead of a national lockdown announced by the Government the following week (Public Health, England, 2020).

At this point, students had been briefed on their individual projects and were beginning these. The lack of new content delivery at this stage in the module was advantageous for the sudden shift to remote delivery. Students were guided through installing software (e.g. Python) on their own machines and so enabled to work on their projects at their own pace. The module staff member was available remotely via email, live text chat or video call to offer feedback and support at student request. A requirement of the project is that students provide progress updates during in-class meetings in weeks two and five of the six-week project, and these meetings were changed to email updates. The first of these updates was received on 20th March 2020, with students who did not participate followed up to see if they needed assistance engaging with the module remotely. Nearly all students submitted projects and marks were comparable to the previous year.

4. Change in delivery for 2020/21

4.1. Principles of teaching approach

Teaching began for the 2020/21 academic year with a new cohort of 18 students taught wholly remotely for this module.

It did not seem sensible to replace the classes which took place in person with pre-recorded videos since this would mean a significant shift away from student-centred activity towards staff content delivery via information transmission. Neither did it seem sensible to arrange classes as shared online video calls, because this would require students to maintain a live connection over the internet (not straightforward for all) and seemed to offer little benefit since class time is mostly spent on individual work. Instead, the module was delivered via a flexible, student-centred remote learning approach.

Notes were provided via the web in the usual way. Classes took place asynchronously, with students working through notes and exercises and asking for help as needed via online tools. Students were given flexibility over when they participated in this, though a weekly class timeslot was maintained where module staff were particularly looking to provide support for the module. Students were encouraged to write with questions, asking for a short video call if needed.

The asynchronous nature of the learning meant that part of the formative assessment process would be missing. The students would still interact with the computer by writing code and could seek help from module staff if required, though note that the barrier to seeking help is raised – writing an email rather than calling someone from across the room for a chat – and some students are reluctant to begin a formal interaction with staff. The missing part is that staff do not have the opportunity to casually observe the student and informally enquire about their progress. Consequently, a formal system of formative assessment was implemented. Along with the usual notes and exercises, a weekly formative task was set. The aim was to enable staff to view students' code, to assess whether students had understood the week's content and provide a context for a feedback discussion in the case where they had not. This was designed to more effectively prompt formative interaction in remote learning circumstances.

4.2. Practicalities

A system was required for students to receive task instructions and submit their work, and GitHub Classroom (https://classroom.github.com/) was selected for this. GitHub is a website that allows storage and sharing of code based on the version-control system `git`. GitHub Classroom is an education-focused tool that allows a task to be set and distributed to students, and their submissions to be collected.

The use of GitHub for this purpose is not inappropriate for this kind of module, indeed it had been recommended by Jones Megeney (2020) for teaching programming in undergraduate mathematics, albeit in a group work setting. It had not been implemented previously as it seemed to be an extra complication that is not core to the needs of undergraduate mathematics students. The opportunity to use the system to submit code for feedback was felt to sufficiently outweigh the distraction of having to learn to use it.

The first class ran as a live video call, and the focus was the delivery of information on how the class would work and student activities of installing Python and submitting a basic task via GitHub. These activities were necessary for engagement with the rest of the module, so the live video call was felt necessary to provide support and avoid a barrier to engagement. Students worked in individual break-out rooms and once each student had completed both tasks, they left the call.

Thereafter, teaching took this form:

- At the end of each class, the notes for the following week were released and students were emailed. These emails highlighted the key learning objectives of the next week's content, provided any relevant instructions and included any other module news. Emails were sent a week in advance of the class time so that students had the flexibility to work through the next topic at their convenience.
- At the start of the next class, this email was resent as a reminder that the class time was now live for those students who chose to work during the designated time.
- During class, answers and feedback were sent first to students who had already asked questions or submitted work earlier in the week, and then to students working during

(1) Write a function called `closure_date` which works for dates in the current academic year. It should take in a date formatted as YYYY-MM-DD and returns `True` if that date is a University closure day, and `False` otherwise.

(2)
   (a) Import `math` and use it to calculate the value of $\cos(5\pi + 2)$.
   (b) Import `emath` and use it in a loop to print the square roots of $1 + ki$ for $k = 1, \ldots, 5$.
   (c) Import `sympy` and use it to print the 2nd derivative of $\sin(\mathrm{e}^x)$.
   (d) Use `sympy` to invert a 3×3 matrix of your choosing.

(3) Make a new Excel Macro-enabled worksheet and save it in this folder. Create a UserForm that collects the following information from the user.
   - Name (using a TextBox);
   - Favourite colour (using a ComboBox);
   - Whether you have read 'Hello World: Being Human in the Age of Algorithms' by Hannah Fry (using 'yes'/'no' OptionButtons).

   Use a CommandButton to trigger inputting the data.
   - If any items are left blank, your program should use a MsgBox to ask the user to fill these in and not store the data.
   - If all items are present, your program should store the submitted information in a table in the Excel spreadsheet and hide the UserForm.

(4) Using the MathsJam database file, write an SQL query that identifies the tweets in the database that are by me which mention the word 'baking'. Submit to this repository a plain text file `results.txt` which contains
   - The SQL query that uniquely identified these tweets.
   - The `tweet_text` of the matched messages.

(5) Write a Python script that outputs an `.html` file containing a HTML+MathJax document which displays an integral with at least one randomised constant.

Figure 1. Examples of formative tasks set.

the class time as the need arose. Students were asked to submit their formative work by one hour after the end of the class, though this deadline was not enforced.

For questions, students were encouraged to email or to make use of the GitHub Gist system (https://gist.github.com/), which allows the sending of snippets of program code without this being garbled by an email system.

A selection of the sixteen tasks set during the teaching year is presented in Figure 1.

Work submitted via GitHub is timestamped and presented to module staff, and feedback can be attached to code that has been submitted. An example of the feedback given via GitHub is shown in Figure 2. Fiksel et al. (2019) describe a similar use of GitHub Classroom in the teaching of computational statistics, but with feedback provided by inline comments in the students' submitted code.

Delivery was to a small group of students. The approach to formative assessment via GitHub would scale with increased staffing in response to increasing student size, since it is based on staff writing individual responses to student submissions. However, GitHub Classroom provides some automated marking features (called 'autograding' by Jones, 2020) which may be advantageously explored for large groups; here we felt numbers were such

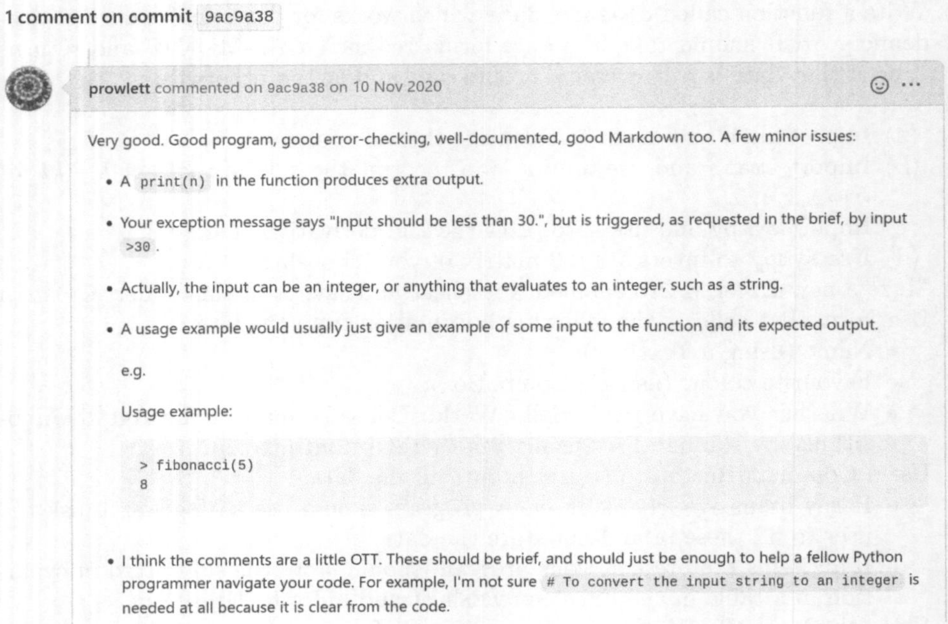

Figure 2. Example feedback on code given via GitHub.

that the extra development time in an already-busy year did not outweigh the reduced marking time.

5. Student feedback

An attempt was made to collect student views on the module via a questionnaire after teaching had finished and the projects were submitted. This was approved by the Sheffield Hallam University research ethics process (approval no. ER33259510) and distributed to students, but only four students responded and none filled in the free-text questions, limiting the possible analysis. These responses, along with reflection on more informal feedback, are presented in this section. Other sources of feedback that feed into module staff reflections include emails, comments in response to feedback on GitHub, a course-level personal development portfolio and formal course student feedback meetings.

5.1. Live delivery

Survey responses were mixed about the need for live delivery. One student agreed there should be live delivery of taught content, including even if this meant staff time was used up and formative feedback could not be given; though this same student strongly agreed that submitting weekly work had aided their learning. The other students disagreed that live delivery was needed. No negative feedback was received through formal course feedback processes or via other routes about the delivery approach for this module.

Survey respondents agreed that a weekly live drop-in video call should have been offered to answer questions. This is worth considering, though note that a formal online ‘office hour' video call was offered for each staff member during the second half of the year, with zero take-ups from students.

5.2. *Formative weekly work*

Most survey respondents reported submitting most weekly work and generally finding this straightforward to do. Survey respondents did not feel they dropped behind because of the self-paced nature of the work. In fact, almost all students submitted work regularly at the start of the year, though the numbers submitting formative work decreased as the module progressed. It would be interesting to know why, though this does not seem out of line with the usual experience elsewhere of reduced take-up of formative opportunities as the pressure of summative work in other modules increases. Note also that students commented informally about the tasks getting harder through each topic, which may explain some disengagement.

Survey respondents agreed that submitting work for feedback aided their learning and felt they were getting enough feedback. Informal student comments throughout the year were similarly positive on this point.

5.3. *Alternative delivery circumstances*

The survey asked students to imagine taking the module in different contexts: (a) in-person teaching with social distancing requirements (sitting apart while wearing masks); (b) no restrictions at all. Two students felt classes should take place in either circumstance, though interestingly one of these had disagreed that live online delivery should form part of the module this year. A third respondent agreed classes should take place only if there were no restrictions at all. A fourth student disagreed there was a need for in-person classes in either circumstance.

6. Discussion/reflection

The redesign of the module delivery involved shifting staff time from live delivery and informal interaction designed to prompt a formative interaction to a formal formative assessment process. The idea for this change arose from concern that the loss of informal in-class interaction would mean that formative interaction was not effectively prompted, but as this resulted in more student work being thoroughly reviewed flaws in the informal system became apparent. It was clear from reviewing work that students often think they have completed a task successfully when in fact they have made some mistake perhaps undetected by the computer response. Consequently, it appears informal questioning in class about the outcome of exercises is not a reliable method of identifying learning misconceptions or development needs when compared to a review of student work. In some ways, this seems surprising, given that program code will give its own feedback in terms of errors, warnings and incorrect output. We theorize that early learners in programming topics might lack the sophistication to identify subtle issues with the code they have written.

It seems, therefore, that answering questions asynchronously and providing detailed feedback on code is an effective use of staff time for this delivery. It is important to note that not all students engaged with formative work consistently throughout the year, raising issues of disadvantage for the less-engaged. Although it is based on a small sample size, it is interesting to note that not all students feel that classes are needed for this module even if there are no restrictions. However, it is important to know that some would rather be in class even if the circumstances were not much improved.

7. Takeaways

Novice programmers do not appear to be able to successfully detect subtle issues in the code they have written, meaning that simple exercises do not necessarily provide an effective prompt for formative interaction. Therefore formal submission of work for formative feedback is recommended as an effective instructional strategy for programming, even if this means diverting staff time to providing feedback on formative tasks.

It is not clear that an introductory programming module need necessarily take place with in-person classes, even in a teaching situation where classes are usually in person.

Disclosure statement

No potential conflict of interest was reported by the authors.

ORCID

Peter Rowlett http://orcid.org/0000-0003-1917-7458
Alexander S. Corner http://orcid.org/0000-0001-6222-3443

References

Fiksel J., Jager L. R., Hardin J. S., & Taub M. A. (2019). Using GitHub classroom to teach statistics. *Journal of Statistics Education*, *27*(2), 110–119. https://doi.org/10.1080/10691898.2019.1617089

Hattie J., & Timperley H. (2007). The power of feedback. *Review of Educational Research*, *77*(1), 81–112. https://doi.org/10.3102/003465430298487

Jones A. (2020). Introducing autograding for GitHub classroom and the GitHub teacher toolbox.

Jones M. M., & Megeney A. (2020). Programming in groups: developing industry-facing software development skills in the undergraduate mathematics curriculum. *MSOR Connections*, *18*(2), 18–24. https://doi.org/10.21100/msor.v18i2

Public Health, England (2020). The health protection (Coronavirus, Restrictions) (England) Regulations 2020 (SI 2020/350).

Rowlett P. (2020). Programming as a mathematical activity. *MSOR Connections*, *18*(2), 13–17. https://doi.org/10.21100/msor.v18i2

Schoenfeld A. H. (2015). Summative and formative assessments in mathematics supporting the goals of the common core standards. *Theory Into Practice*, *54*(3), 183–194. https://doi.org/10.1080/00405841.2015.1044346

Taras M. (2005). Assessment – summative and formative – some theoretical reflections. *British Journal of Educational Studies*, *53*(4), 466–478. https://doi.org/10.1111/j.1467-8527.2005.00307.x

Don't throw the student out with the bathwater: online assessment strategies your class won't hate

Stuart Johnson, John Maclean, Raymond F. Vozzo, Adrian Koerber and Melissa A. Humphries

ABSTRACT

STEM educators have taken on a new and radically different reality: that students will have access to online tools and online knowledge as they undergo assessment. Students may look up facts that are easily available online, employ online tools and also employ paid services to cheat. The first wave of novel assessment strategies devised in response to this new reality was online invigilation measures that were often either ineffective or widely hated by students. This paper considers two frameworks for effective online assessment in the modern era: a framework for question design in STEM and a framework for the structure of assessment. The framework for question design discusses methods to discourage the use of online tools and calculators to simply obtain answers and methods to detect contract cheating. The framework for assessment structure discourages cheating while managing to increase student satisfaction and engagement with content: we present a case study from a cohort of 113 first-year university students of adapted assessment strategies with no identified paid cheating, well-differentiated student results and large-scale positive student feedback. Our key contribution is to identify simple and effective steps to assess students fairly in an online environment, without empowering or motivating them to cheat.

1. Introduction

University STEM courses have traditionally placed much of the weighting of the course grade on one or two major exams. This is largely motivated by the nature of closed-book, invigilated exams: instructors can be more-or-less certain that the work being assessed has been produced by the student whose grade is being determined. One of the biggest impacts that COVID-19 has had on STEM courses at university has been the loss of this method of assessment. The standard method for assessment in 2020 (and beyond) at the University of Adelaide has been *un*invigilated, open-book, online examinations that present challenges in ensuring that students are doing their own work,[1] and in writing assessments that are not trivial to answer with online tools and resources.

The shift to online assessment has measurably worsened some components of academic honesty: contract cheating has increased. The usage of Chegg, for example, increased threefold from March 2019 to March 2020[2] (Lancaster & Cotarlan, 2021). Online proctoring has been taken up as a solution and has drastically increased since 2019 (Kimmons & Veletsianos, 2021). Although online proctoring may seem like the perfect solution, there is some evidence that assessment design is more significant than proctoring (Mellar et al., 2018). Additionally, issues have been raised with online proctoring; students often find it invasive,[3] and there are concerns about data privacy (Flaherty, 2021; Stewart, 2020).

However, cheating was not always a higher risk in online assessment: late 2000s research reported in several cases that the prevalence of cheating in online assessment was roughly equal to cheating in face to face assessment (Watson & Sottile, 2010). A meta-analysis of 79 papers studying the motivations behind academic honesty and academic dishonesty finds that students who perceive assessment structure as based on developing mastery of particular skills, and who are convinced that the course is useful, are less likely to cheat (Krou et al., 2021). This broader perspective on student honesty reveals an opportunity to reverse the arms race, of student cheating vs attempts to eliminate the possibility, described above.

This manuscript is primarily concerned with the methods and philosophy of non-intrusive assessment strategies that empower educators to assess student learning. We present a suite of solutions, with examples, to redesign course structure and/or content aimed at maximizing student learning and course satisfaction, decreasing cheating and maintaining differentiability of results.

2. Background

Two case studies were carried out at the University of Adelaide. In the first study, the types of assessment question asked in three first-year mathematics courses were altered to preserve their ability to test student knowledge in an open-book format. In the second study, of a first-year statistics course, the structure of assessment was altered with the same goal. We report the goals for alterations to the mathematics courses in Section 2.1 and the alterations themselves – with diverse examples – in Section 3.1. The goals for the statistics case study are reported in Section 2.2 and the alterations described in Section 3.2. Results of these alterations are the topic of Section 4, and we conclude in Section 5.

2.1. Question structure

The first-year mathematics courses required significant redesign of exam question style to adapt to online exams. In addition to being able to access assignment help sites, students would also have access to their course notes and to online mathematical solvers. A review of a 2019 first-year mathematics exam revealed that 65% of the marks could be obtained using WolframAlpha alone.

Open resource exams in mathematics appear to be rare prior to 2020. Examples include service teaching for specific applications (Nazim Khan, 2015) or involve further forms of assessment such as an oral exam (Cnop & Grandsard, 1994). Open resource exams with sophisticated computer algebra capabilities available to students were studied when CAS calculators first became available, see Pountney et al. (2002), but their use does not seem to have been widely adopted. A more recent study, Maciejewski (2021), was open resource but employed invigilation to prevent unauthorized communication.

Section 3.1 reports redesigned question strategies employed – on short notice – in 2020 to mitigate the problems with open resource exams.

2.2. Assessment structure

The first-year statistics course changed from a single, 70% exam to a series of assessments that were designed to circumvent the issues itemized above. Using principles of continual assessment (see recommendations, and warnings, in Hernández (2012) and Harland et al. (2015)) to increase engagement with the course and retention of concepts, the course assessment was split into two broad categories:

(1) Low-weight, collaborative assessment and
(2) Higher-weight, independent assessment.

The low-weight assessment was used at regular intervals to encourage both engagement with the course and engagement with their peers. This aimed to build a sense of community and extend learning by encouraging safe and careful exploration of difficult concepts. To differentiate marks on individual understanding and application, higher-weight assessment was used. These were staggered across the semester to provide regular points of consolidation of course concepts and limited in time to reduce the opportunity to cheat (Michael & Williams, 2013).

3. Methods

3.1. Question design for assessing students with access to online mathematical solvers

Below we will illustrate some of the strategies used in the mathematics courses, referring to examples of questions used prior to and during 2020 which are given in Appendix. Crucially, the questions test similar skills and knowledge and are of a similar difficulty.

3.1.1. Graphical presentation

As students would have graphing tools available to them, we switched from asking for graphs or answers which would be found by graphing, to testing the same skills by asking students to interpret graphs.

An example showing an old and new question created with this approach is given in Figure A1.

Figure A2 shows a question on qualitative methods for differential equations where the function is given graphically rather than as a formula so that they cannot simply use online tools to numerically solve the equation.

3.1.2. Partial working for questions that cannot be solved using online tools

Typically, an exam question that is too complicated for an online tool is also too complicated for a student. However, it is possible to test very specific aspects of a method on such a problem. For example, Figure A3 demonstrates an online-appropriate integration by substitution on an integral that is not possible to calculate in closed form. Thus an online calculator will give up and give a numerical answer, yet it is possible for a student to demonstrate basic knowledge of the method of substitution.

3.1.3. Giving incomplete information

Figure A4 shows that the methods of solving a DE may be assessed without giving an equation which can be solved online by instead giving the integrating factor, making the question a straightforward test of the method for those who understand, yet impossible using an online calculator. Figure A5 shows contrasting approaches in testing understanding of results on eigenvectors and diagonalization, without simply giving an explicit matrix to diagonalize.

3.1.4. Working backwards

For some problems it is possible to start with the solution and ask for the problem, for example asking for a differential equation with a given solution, see Figure A6.

3.1.5. Asking for working or specific methods

Figure A7 shows polynomial division questions from 2019 and 2020, respectively. Since this is very standard and can easily be done by online tools, asking for a specific layout can help to determine if the student has studied the course materials.

3.1.6. Question design to deter and detect contract cheating

The first-year mathematics exams included both a written and a computer entered (Mobius) section.

The chief strategies were:

Use of distinctive wording. Many common questions are online already, and may already be solved there, so the detection of cheating is much harder. Including local references can be effective: for example instead of 'A factory produces …' we used 'A South Australian factory produces …'.

Referencing specific details of the course. To reduce the ability to get answers from contract cheating websites specific course references were used, for example 'Use Theorem 15.3 to explain why …' or 'You must set out you calculations in the way used in lectures'.

Randomization. Using Mobius we have been able to randomize questions to the extent that students receive a unique version. Any upload to a website can then be uniquely matched to a student.

The question in Figure A8 has randomized matrix entries, including some that do not affect the solution but aid in randomization for detection purposes.

The question in Figure A9 is more theoretical (which can be harder to randomize), but in this case the parent question contains over 200 million versions. We randomized the statements given and the parameters contained within them. In order to efficiently match any question on Chegg to the student that had posted it, we included hidden text in the question that lists the parameter values in html source. When a copy of the question was posted online, we deduced the list from the parameter values in the question, and then used the list to immediately find the matching exam question and prove that only one student had been given that precise question.

3.2. Assessment design

As an alternative to retaining a single, high weight assessment, a first-year statistics course of 113 first-year Mathematics students sought to address assessment issue by changing the assessment structure.

The course had previously had five assignments and a 70% final invigilated exam. This was adapted to embrace continual assessment; utilizing quizzes, statistical written reports and smaller, regular mathematics tests across the semester. Assessment was split into low-weight, collaborative components that formed 30% of the total grade while the remaining majority of students marks came from independent assessment pieces; major quizzes, statistical reports and mathematics tests.

3.2.1. Low-weight, collaborative assessment

There were two low-weight assessment styles used throughout the course, aimed at encouraging engagement with the material, providing opportunities to practice learning, and increasing collaboration, fostering a sense of community. The first of these styles was weekly, low-weight quizzes to incentivize regular engagement and provide a safe environment for testing learning. Timing was generous to enable students time to look through their notes or collaborate to find answers. The questions were heavily randomized using the `exams` package in `R`, so that even with collaboration, students would need to apply the methods learned to their own scenario.

The second low-weight assessment was special assignments that students were encouraged to do together. Due at three points across semester, the assignments included questions that most students would be incapable of answering alone. The questions were designed to encourage unique thought and exploration and required collaboration and research to achieve a result. These special assignments led students through proofs and examples that underpin some of the more difficult concepts in the course, providing depth of understanding.

3.2.2. High-weight, independent assessment

Major quizzes, similar in content to the low-weight quizzes, were held at mid and end of semester. In order to reduce collaboration, collusion and contract cheating, the major quizzes were randomized using `R/exams` and had a reduced available completion time relative to the number of questions being asked (Michael & Williams, 2013).

Restricted timing was also utilized for the three mathematics tests, run in class, across the semester. These tests were semi-invigilated, open-book tests that ran for both face to face and online only cohorts. For the online cohorts, the test was completed via zoom. Students showed their student ID and were requested to keep their video on during the test. The cohort was split into break out rooms of no more than 10 students with a tutor monitoring each break out room. As the tests were run in class, students had 50 minutes to complete them, reducing their ability to apply to contract cheating sites for answers. Students were primed to expect questions based on the topic videos from the course, increasing the likelihood that students went first to the course notes or videos for information, than using other online sources.

The final high-weight assessment pieces were three written reports and a prac test, based on statistical analysis completed in prac classes. Reports consisted of a technical section, detailing their statistical analysis, and an executive summary to communicate their findings. As this was a worded assignment, it could be run through a plagiarism checker, encouraging students to complete their own report even if they shared analyses. To test for students' capability in running analyses, the report was supplemented with a final prac test (in class). Students were given a question related to their current analysis to answer and incorporate into their report within 50 minutes. This tested computational ability in a time pressured environment but with data, and a project, they were familiar with. Using this technique reduces the desire to cheat by building on their current work (Lancaster & Clarke, 2016).

4. Results

We report three areas:

(1) Whether the course was able to differentiate ability between students,
(2) student satisfaction,
(3) the ability to detect cheating.

4.1. *Ability to differentiate*

One of the measures of success for a course is its ability to differentiate between the ability and understanding of students. Table 1 shows the mean and standard deviation for the 2020 courses discussed in this manuscript, along with their 2019 counterparts for comparison. A good distribution of grades was maintained across all courses, evidenced by the standard deviations around 23 marks.

Comparing the results in 2020 to those in 2019 gives a measure of consistency of outcomes across the 2 years. Using two-sample t-tests (equal variance not assumed), we found no significant difference between the 2019 and 2020 grades for statistics ($t(207) = 0.25$, $p = .803$), or between any of the reconfigured exams for Introductory Mathematics ($t(160) = 0.45$, $p = .656$), Intermediate Mathematics ($t(618) = 1.35$, $p = .178$) or Advanced Mathematics ($t(976) = 0.44, p = .658$). In these calculations, a value of p larger than 0.05 indicates that there is no observable difference between the years, and hence there is no evidence that the changes to assessment structure changed student outcomes.

Table 1. Means, standard deviations (sd) and class sizes (n) for the 2020 statistics course and the 2020 mathematics exams discussed in this paper, alongside their 2019 counterparts. All students with valid grades were counted. It is clear that there is a good distribution of grades across all courses and that there is little difference in students outcomes between 2019 and 2020, despite the stresses associated with lockdown and changing assessment.

	2019		2020	
Subject	Mean *(sd)*	*n*	Mean *(sd)*	*n*
Statistics Exam	69.9 *(21.16)*	106	70.6 *(22.19)*	104
Introductory Mathematics	48.7 *(22.67)*	88	50.2 *(22.10)*	76
Intermediate Mathematics	59.1 *(17.29)*	351	57.1 *(21.10)*	320
Advanced Mathematics	60.2 *(20.19)*	475	60.7 *(18.71)*	555

In other words, the radical alterations in assessment delivery did not alter the difficulty of the assessment or the differentiability of the student cohort.

4.2. Student feedback

At the University of Adelaide, students are offered the opportunity to complete a student evaluation questionnaire at the completion of each course. Questionnaire responses are recorded before the exams are taken; student feedback is therefore irrelevant when considering assessment design in the mathematics courses, for which the design of exams was the key focus. However, the continual assessment structure of the statistics course, laid out in Section 3.2, meant that all assessment had been concluded by the time of the questionnaire, and we therefore report on student feedback.

For the first-year statistics course, 95% of 38 respondents broadly agreed with the statement, '*The assessment tasks in this course help me learn*'. With a maximum score of 7, this course scored a mean of 6.3 and median of 7. This was a higher mean than both the School of Mathematical sciences (6.0) and the University in general (5.8). In addition to the Likert responses, 28 individuals gave positive written feedback on the course and, of those, 22 ($\sim$ 79%) mentioned their appreciation of the adapted assessment.

As an example of the written student feedback, one student said:

> The ability to have three tests throughout the semester allowed for me to grasp concepts better due to the constant revision as well as taking off some major stress of an end of year exam. Because the weighting of each of these is still high, they carried significance but the knowledge that bombing one didn't mean you would necessarily fail was a weight off of my shoulders.

Other comments included an appreciation that they were '*...forced to keep up with the work as we go*' and a student, '*...was able to really absorb the material as opposed to cramming.*', providing evidence that the goals of the changed assessment were met in this dimension.

4.3. Detection of integrity breaches

The mathematics course sought to reduce the opportunity to cheat as discussed in Section 3.1. As discussed in Section 3.1.6, moreover, the mathematics course included randomization of questions, and instances of contract cheating could be matched to the precise student who had uploaded the question.

The statistics course included some of the strategies described in Section 3.1 but, as described in Section 3.2, extensively altered assessment design to focus on repeated, short-duration assessment. For this course, Turnitin found no examples of plagiarism across the semester. Regular checks across known contract cheating sites returned only one posted question (of one special assignment – low-weight, group work) that went unanswered.

5. Conclusion

We have discussed two complementary approaches to course (re)design for STEM educators in a world of online assessment. Both approaches were motivated by the need to alter assessments without dramatically changing the difficulty of questions or the skills being tested.

It is fair to mention here that it is not always a realistic goal to eliminate cheating in an online course. Primary motivations for students to cheat relate to extrinsic and intrinsic motivations and goal structures (see Krou et al. (2021) for a meta-analysis), and some of these factors will always apply to certain student cohorts; for example, Flegg et al. (2012) and Harris et al. (2015) discuss perspectives of engineering students regarding mathematics. Nonetheless, we highlight the following key points.

Takeaways

- When carefully implemented, continual assessment across the semester results in good differentiability of student performance along with increased student satisfaction.
- With careful question design, it is largely possible to redesign mathematics exams to test course material in a manner suitable for the open resource environment.

Data sharing: The University of Adelaide does not require specific ethics approval to share the pooled student results, or anonymous student feedback, reported in this paper.

Notes

1. Seaton (2019) discusses the modern forms of academic misconduct, distinguishing plagiarism, copying, collusion and contract cheating
2. While Chegg markets itself as a site for 'homework help', it is commonly used by students to obtain test or assignment solutions.
3. One report finds 17 major petitions against their use (Kelley, 2020)

Disclosure statement

No potential conflict of interest was reported by the author(s).

ORCID

John Maclean http://orcid.org/0000-0002-5533-0838
Melissa A. Humphries http://orcid.org/0000-0002-0473-7611

References

Cnop I., & Grandsard F. (1994). An open-book exam for non-mathematics majors. International Journal of Mathematical Education in Science and Technology, 25(1), 125–130. https://doi.org/10.1080/0020739940250116

Flaherty C. (2021). Online proctoring is surging during COVID-19. https://www.insidehighered.com/news/2020/05/11/online-proctoring-surging-during-covid-19.

Flegg J., Mallet D., & Lupton M. (2012, September). Students' perceptions of the relevance of mathematics in engineering. International Journal of Mathematical Education in Science and Technology, 43(6), 717–732. https://doi.org/10.1080/0020739X.2011.644333

Harland T., McLean A., Wass R., Miller E., & Sim K. N. (2015, May). An assessment arms race and its fallout: High-stakes grading and the case for slow scholarship. Assessment & Evaluation in Higher Education, 40(4), 528–541. https://doi.org/10.1080/02602938.2014.931927

Harris D., Black L., Hernandez-Martinez P., Pepin B., Williams J., and with the TransMaths Team (2015, April). Mathematics and its value for engineering students: What are the implications for teaching? International Journal of Mathematical Education in Science and Technology, 46(3), 321–336. https://doi.org/10.1080/0020739X.2014.979893

Hernández R. (2012, October). Does continuous assessment in higher education support student learning? Higher Education, 64(4), 489–502. https://doi.org/10.1007/s10734-012-9506-7

Kelley J. (2020, September). Students are pushing back against proctoring surveillance apps.

Kimmons R., & Veletsianos G. (2021). Proctoring software in higher ed: prevalence and patterns. *EDUCAUSE Review*. https://er.educause.edu/articles/2021/2/proctoring-software-in-higher-ed-prevalence-and-patterns.

Krou M. R., Fong C. J., & Hoff M. A. (2021, June). Achievement motivation and academic dishonesty: A meta-analytic investigation. Educational Psychology Review, 33(2), 427–458. https://doi.org/10.1007/s10648-020-09557-7

Lancaster T., & Clarke R. (2016). *Contract cheating: The outsourcing of assessed student work*. Springer Singapore. pp. 639–654.

Lancaster T., & Cotarlan C. (2021). Contract cheating by STEM students through a file sharing website: a Covid-19 pandemic perspective. International Journal for Educational Integrity, 17(1), 1–16. https://doi.org/10.1007/s40979-021-00070-0

Maciejewski W. (2021). Let your students cheat on exams. PRIMUS, 31(6), 685–697. https://doi.org/10.1080/10511970.2019.1705450

Mellar H., Peytcheva-Forsyth R., Kocdar S., Karadeniz A., & Yovkova B. (2018, December). Addressing cheating in e-assessment using student authentication and authorship checking systems: teachers' perspectives. International Journal for Educational Integrity, 14(1), 1–21. https://doi.org/10.1007/s40979-018-0025-x

Michael T., & Williams M. (2013, October). Student equity: discouraging cheating in online courses. Administrative Issues Journal, 3(2), 30. https://doi.org/10.5929/2013.3.2.8

Nazim Khan R. (2015). Assessments: an open and closed case. International Journal of Mathematical Education in Science and Technology, 46(7), 1061–1074. https://doi.org/10.1080/0020739X.2015.1076534

Pountney D., Leinbach C., & Etchells T. (2002). The issue of appropriate assessment in the presence of a CAS. International Journal of Mathematical Education in Science and Technology, 33(1), 15–36. https://doi.org/10.1080/00207390110086616

Seaton K. A. (2019). Laying groundwork for an understanding of academic integrity in mathematics tasks. International Journal of Mathematical Education in Science and Technology, 50(7), 1063–1072. https://doi.org/10.1080/0020739X.2019.1640399

Stewart B. (2020). Why higher ed needs data ethics | inside higher ed.

Watson G. R., & Sottile J. (2010). Cheating in the digital age: Do students cheat more in online courses? Online Journal of Distance Learning Administration, 13(1), 14–23. https://eric.ed.gov/?id=EJ877536

Appendix. Example exam questions

Let $f(x) = \|x - 2\|$ with domain $x \in [-2, 4]$. (i) Sketch the graph of f.	(a) Let $f(x) : [0, 5] \to \mathbb{R}$ be a function whose graph is shown below 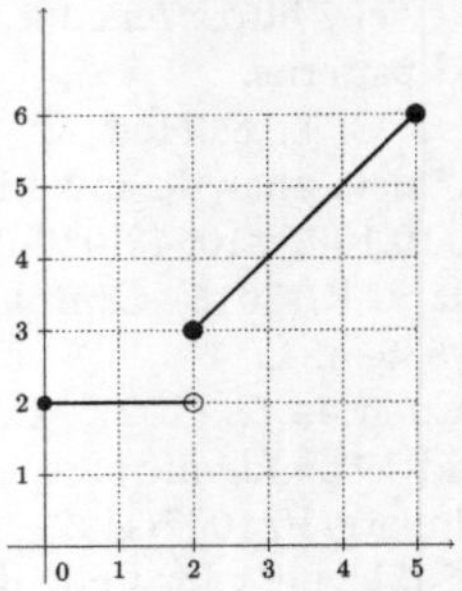 (i) What is $f(2)$? (ii) Is $f(x)$ 1-1? You must give reasons for your answer. (iii) Let $g(x) : [1, 6] \to \mathbb{R}$ be defined by $g(x) = f(x - 1)$. Sketch the graph of $g(x)$. You must label your axes to make the scale clear.

Figure A1. Question on real-valued functions. *Left:* 2019 question, easily solved with diverse online tools. *Right:* 2020 question.

Consider the differential equation

$$\frac{dy}{dt} = f(y)$$

where $f(y)$ is the function with graph

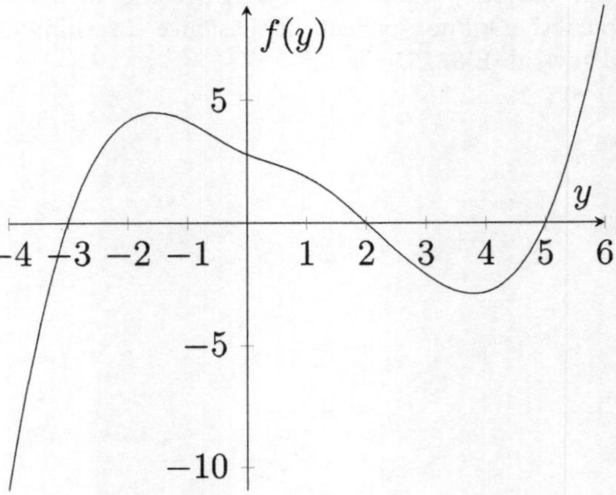

Use a phase-line analysis to determine $\lim_{t \to \infty} y(t)$ if $y(0) = 3$.

Figure A2. 2020 question on differential equations.

Make the substitution $u = 1 + x^2$ to evaluate $\int \frac{x}{1+x^2}\,dx$.	Consider the integral $\int_{\frac{\pi}{2}}^{\pi} \sin(2x)e^{\sin^2(x)}\,dx.$ Use the substitution $u = \sin(x)$ to formulate this as an integral in terms of u. (**Do not** attempt to evaluate the integral.)

Figure A3. Question on integration by substitution. *Left:* 2019 question, easily answered (with working) by WolframAlpha. *Right:* 2020 question.

Solve the separable differential equation $\frac{dy}{dx} = xy^2$, with $y(1) = -1$.	Consider the initial value problem $\frac{dy}{dt} + p(t)y = \cos(e^t), \quad y(\ln(\pi/2)) = 1.$ You are told that an integrating factor for the DE is $\mu(t) = e^t$. Use this information to find the solution of the initial value problem.

Figure A4. Question on differential equations. *Left:* 2019 question, easily answered (with working) by WolframAlpha. *Right:* 2020 question.

A 3×3 matrix B has three eigenvalues $1, 2, 3$. The eigenspaces are $\mathbb{E}_1 = \text{span}\left\{\begin{bmatrix}1\\1\\2\end{bmatrix}\right\}, \quad \mathbb{E}_2 = \text{span}\left\{\begin{bmatrix}1\\0\\-1\end{bmatrix}\right\}, \quad \mathbb{E}_3 = \text{span}\left\{\begin{bmatrix}0\\1\\-2\end{bmatrix}\right\}$ Write down a matrix P such that $P^{-1}BP = \begin{bmatrix}3&0&0\\0&2&0\\0&0&1\end{bmatrix}$.	A 3×3 real symmetric matrix C has two distinct eigenvalues. Suppose one of the eigenvalues is known to be $\lambda = 2$. If $\text{tr}(C) = 7$, then there are two possible values for the second eigenvalue. Find them, giving reasons.

Figure A5. Two examples of incomplete information. *Left:* on diagonalization of a matrix. *Right:* on eigenvalue calculations.

Find a second order constant coefficient homogenous differential equation whose general solution is

$$y_h(t) = Ae^t\cos(2t) + Be^t\sin(2t)$$

where A and B are constants. Show your working.

Figure A6. 2020 question asking for a DE rather than the solution.

Let $p(x) = 2x^3 + x - 1$. 10(a). Use polynomial long division to divide $p(x)$ by $x + 2$.	Use polynomial long division to divide $x^3 + x + 7$ by $x^2 - 2x$, to find the quotient and remainder. You must set out your calculation in the way used in lectures.

Figure A7. Polynomial division. *Left:* 2019 question, for which many correct solutions can be searched online. *Right:* 2020 question.

Let $A = \begin{bmatrix} 1 & 20 & -19 & 11 \\ 0 & 0 & 0 & -38 \\ 0 & 0 & 0 & -19 \\ 0 & 0 & 0 & 0 \end{bmatrix}$.

Select from the following all vectors which are in $\mathbf{Nul}(A)$:

☐ $(-1,-1,1,0)^t$

☐ $(-1,1,1,0)^t$

☐ $(0,38,40,0)^t$

☐ $(-11,0,0,1)^t$

Figure A8. A randomized Mobius question with over 14 million versions (see discussion in Section 3.1.6).

Select from the following statements about eigenvalues, eigenvectors and diagonalisation, all of those which are **TRUE**

☐ All 4×4 matrices with characteristic polynomial $p(\lambda) = (\lambda - 1)(\lambda - 5)(\lambda - 8)^2$ are diagonalisable

☐ If a diagonalisable matrix has characteristic polynomial $p(\lambda) = (\lambda - 1)^2(\lambda - 2)(\lambda + 3)^2(\lambda + 5)$, then given non-zerc vectors $\mathbf{u}, \mathbf{v} \in \mathrm{E}_2$, the set $\{\mathbf{u}, \mathbf{v}\}$ must be linearly dependent

☐ Every 4×4 matrix with characteristic polynomial $p(\lambda) = (\lambda - 12)^4$ is diagonalisable

☐ If A is a 6×6 matrix with characteristic polynomial $p(\lambda) = \lambda^2(\lambda - 7)^4$ then $\det(A) = 0$

Figure A9. A randomized Mobius question addressing theory.

OPEN ACCESS

Forced to improve: open book and open internet assessment in vector calculus

Tracy S. Craig and Tugce Akkaya

ABSTRACT

Calculus is traditionally tested, at least in part, by a closed book written examination taken under secure conditions. In March 2020, two-thirds of the way through a vector calculus course, our institution went into lockdown and all teaching and assessment had to shift online. While some teachers insisted that remotely written non-proctored examinations remain closed book, we accepted that this would result in inequity, with some students certainly choosing to flout that rule should it be imposed, to their advantage over their classmates. Driven by circumstance, we were forced to design a vector calculus examination that would be written open book and 'open internet' with access to online calculators and computer graphing packages. Proficiency at vector calculus is exhibited far more in the interpretation of information and effective decision-making than in the computational solving of integrals. In this article, we shall present and discuss our experiences, address the challenge of non-proctored examinations and argue for the merit of open book examinations in an otherwise standard vector calculus course.

1. Introduction

Assessment drives student learning and informs the student of what the teacher values (Iannone, 2020). If what we value is higher-order cognitive skills and conceptual understanding, then our assessment needs to make these priorities clear. Mathematics service courses taught to students who need to use mathematics as one of many tools in engineering or physics applications need to foster the problem-solving approaches valued by those disciplines. Thoughtfully designed curricula aim to assess conceptual understanding and help develop mathematical thinking, but it is easy in the absence of a call for change to get stuck in a rut where assessment has become routine and procedural. The circumstances of 2020 forced many of us to reconsider the underpinnings of our teaching and assessment and to refresh outdated systems. We chose 'decision making' as a core skill in vector calculus problems, such as using informed judgement to choose the right mathematical tool for the task at hand (Pepper et al., 2012) and recognizing available options, and hence chose to design our open book non-proctored tests about that core principle. Decisions related to, for example, conservative vector fields and choice of theorem are the key to proficiency in vector calculus. Adjusting our assessment to this new mode was achieved by presenting information in diagrams requiring interpretation, presenting problems without instructions on how to solve them and hence encouraging decision making, and focusing on setting up but not evaluating integrals. Having been forced to change our assessment, we have come to welcome the shift in approach and to look ahead to future teaching and learning benefit.

An open book test written remotely and non-proctored is necessarily also 'open internet' with access to many powerful digital tools. Online calculators are useful for solving integrals whose bounds have already been set, even changing the order of integration if necessary, and computer graphing packages can depict surfaces and curves remarkably well. The opportunities for incorporating digital technologies into mathematics classrooms have become more accessible and more powerful with every year. Dimiceli et al. (2010) discuss some useful aspects of Wolfram Alpha; they point out some weaknesses that no longer exist 11 years later or can be compensated for with other software such as GeoGebra. GeoGebra is versatile free graphing software with a growing library of user-designed dynamic tools. Ponce Campuzano et al. (2019) discuss the use of GeoGebra effectively in the context of vector calculus and Caglayan (2016) recommends GeoGebra for the visualization of double integral and Riemann sum concepts.

Closed book, timed tests, usually without any electronic devices, such as calculators, are the norm in mathematics assessment (Iannone, 2020; Trenholm, 2007). Khan (2015) considers the opportunities afforded by open book testing in statistics and mathematics and observes that while open book tests provide an opportunity for assessing higher level understanding, well-designed closed book mathematics tests often do assess for higher level understanding; he concludes that the argument for open book tests in mathematics is unclear. Faced with an unavoidable shift in assessment due to the coronavirus situation, Bailey et al. (2020) concur that 'open book exams are better suited to questions requiring higher order skills' (Bailey et al., 2020, p. 1) and outline ways in which typical closed book questions can be evaluated and modified to suit this mode. They are guided by Smith et al.'s (1996) MATH taxonomy to suggest some practical suggestions not dissimilar to those discussed in this article and elsewhere (Craig, 2021). Iannone (2020) also suggests types of questions suitable for open book testing and goes further to suggest other types of assessment altogether, such as oral tests. In agreement with several of Bailey et al's and Iannone's question design suggestions (Bailey et al., 2020; Iannone, 2020), Jungic (2020) changed his calculus exam to involve application and explanatory questions based on real data on current public health events. In short, the coronavirus situation has forced many of us to change our assessment practices in creative ways which, while uninvited, have proved to be worthwhile and worth maintaining.

2. Context

The vector calculus course we teach every year runs from early February to early April. At our institution, almost all courses are undertaken for one academic quarter. The vector calculus course under discussion is covered in the third quarter. The assessment has been a traditional closed book test completed without a calculator. Two-thirds of the way through the course the university 'locked down' in response to the coronavirus pandemic, meaning that in-person classes were no longer allowed due to health and safety concerns. All teaching had to take place online and assessment possibilities were limited to online or remote options. Our test date was shifted by one week and we had to completely restructure our plans. With few options available to us at the time we opted for an open book, non-proctored test, handwritten at home and submitted as pdf for grading. We could have insisted on the test remaining closed book (as some colleagues in the same situation in other courses did) and could also have insisted on no use of calculators or computers. We predicted that some students would flout that directive, leading to significant inequity and hence we avoided imposing those constraints.

After a process of reviewing multiple old test papers for the course and breaking down all the sorts of questions we could ask in a limited time test, we settled on a list of criteria for question design. These criteria guided us in designing assessment items, requiring decision-making, offering options, providing diagrams instead of asking students to draw them themselves and setting low to no emphasis on actual computational solving of integrals. In 2021, we continued to apply this model and the examples below are drawn from both academic years.

Two separate sets of challenges confronted us in the two vector calculus courses of 2020 and 2021, the first being setting open book, open internet tests and the second being maintaining academic integrity in the remote non-proctored environment. We regard the steps we took to maintain academic integrity as satisfactory; however, discussion of these steps is beyond the scope of this article. Here, we shall address our response to the first challenge.

3. Examples illustrating principles of question design

Four examples have been chosen to illustrate the principles we used to design our open book, open internet tests. Each is contrasted with an assessment item on the same topic, however, not exemplifying the design criteria (chief amongst which is decision-making) underpinning our more recent approach. These assessment items are drawn from vector calculus tests at our institution, both our own course and others'. To prepare the students in 2020 for this sort of question, a selection of past test papers was made available with accompanying descriptions indicating, which questions were good examples of what they could expect in their upcoming test and which questions were bad examples. In 2021, the students had the previous year's test to prepare them and weekly quizzes we designed to provoke a thoughtful approach to vector calculus questions. One quiz question, for example, was very similar to the test question in Figure 1.

The circulation of a vector field is defined as a line integral along a closed curve, being the boundary of the given surface, described in Figure 1. Stokes' theorem allows the line integral to be substituted with a certain surface integral either over the given surface or any other surface sharing that boundary curve. To answer the question in the example in Figure 1 the student needs to know that the surface can be replaced and then needs to determine the boundaries of the four surfaces described. The question is answered most rapidly if the students already know this information, but they also had access to their textbook, all their notes and any graphing packages available.

We can contrast the Stokes' theorem question above to an arguably more standard item such as 'Use Stokes' Theorem to evaluate the work done by the velocity field $\boldsymbol{v} = (y + xz)\boldsymbol{i} + (2x + yz)\boldsymbol{j} + (z^2)\boldsymbol{k}$ moving a particle along C where C is the boundary of

A textbook exercise says

"Find the circulation of **F** around the boundary of that part of the paraboloid $y = 6 - 2x^2 - 2z^2$ above the plane $y = 2$." ["above" means on the side of greater y, not on the side of smaller y.]

We plan to use Stokes' Theorem to solve this exercise, using the formula $\int_C \mathbf{F} \cdot d\mathbf{r} = \iint_S \nabla \times \mathbf{F} \cdot d\mathbf{S}$. Which of the following is an **incorrect** option for choice of S in this context? Motivate your answer clearly and completely. A diagram by itself is not a suitable motivation.

a) $2x^2 + (y-2)^2 + 2z^2 = 4, y \in [0, 2]$
b) $x^2 + y^2 + z^2 = 8, y \in [2, \sqrt{8}]$
c) $2x^2 + y + 2z^2 = 6, y \in [2, 6]$
d) $x^2 + z^2 \leq 2, y = 2$

Figure 1. Decision-making and processing information.

C is the positively oriented graph of $\mathbf{r}(t) = \langle t^3 - 5t, -t^2 \rangle,\ t \in \mathbb{R}$. R is the region bounded by C and $x = 0$ in the fourth quadrant. Express the area of R as a <u>line integral</u> and then solve for the area of R. Show all your working.

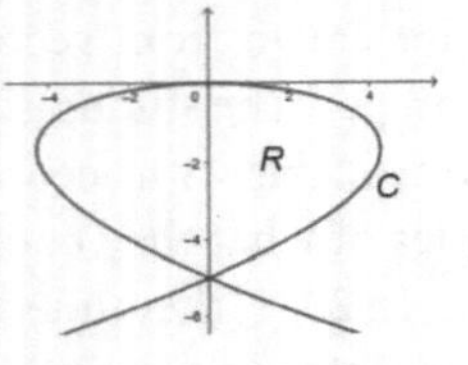

Figure 2. Interpreting information, decision-making, using a diagram.

the hemisphere $z = \sqrt{4 - x^2 - y^2}$.' We are choosing this item as contrast because it also includes the option to exchange the given surface (the hemisphere) with another (a planar disc); however, the failure to notice this option is no impediment to following a standard, memorized, computational algorithm.

Figure 2 shows a region in the xy-plane bounded by a parametrically defined curve. The student is required to calculate the area of half of that bounded region. To solve the example shown in Figure 2, the student needs to know (or know how to look up) that area of a bounded region can be expressed as a line integral along the boundary using Green's theorem. The region described in the example is half of that bounded region. Once having

The density of lamina D at point (x, y) is twice the distance from the point to the y-axis. Write down (but do not evaluate) an iterated double integral to represent the mass of lamina D shown below. Do this first in the order $dxdy$ and then in the order $dydx$. D is bounded by $x = 0, y = 0, y = 2$ and $y = 3 - x^2$.

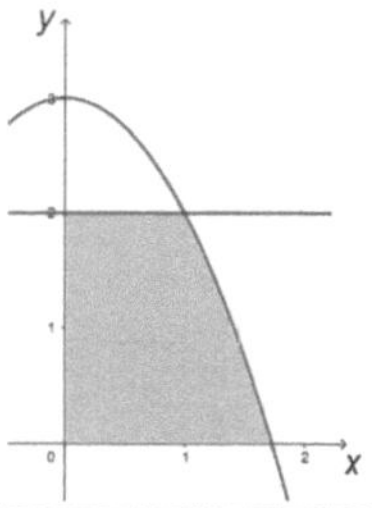

Figure 3. Providing a diagram and not requiring solving an integral.

realized that Green's theorem could be utilized the student would then need to determine the parameter values describing the closed curve. In grading this question, very little weight was given to the computed solution.

This example, requiring the use of Green's theorem to translate the double integral for area into a line integral along the boundary, can be contrasted with items such as 'Use Green's Theorem to calculate the area of the disk D of radius r defined by $x^2 + y^2 \leq r^2$.' This contrasting item stops short of supplying the formula to be used, but otherwise requires very little from the student other than a good memory and computational skill. The bounds of integration are predictable and could be applied by the student with little understanding of what they represent.

Figure 3 shows a region, a lamina, in the xy-plane piecewise bounded by four given curves. The student is required to set up two double integrals, one in each Cartesian order, describing the mass of the lamina. In a standard closed book test, the example in Figure 3 would not include a diagram; drawing a diagram would be part of the expected solution. This question was set in the April 2020 test where the students had no expectation that skill with graphing software might be useful. Requiring that the students include a diagram in their solution would, therefore, result in inequity since two students with equal difficulty in hand drawing such diagrams could fare very differently if one used computer graphing software and the other did not know how to do so. In a significant deviation from how we traditionally asked vector calculus questions we shifted to including diagrams. Representing the low emphasis on computation in our new design scheme this question explicitly did not require any computation. Prior to taking the approach discussed in this article, we have set test questions such as 'Lamina D is bounded by $x = 0, y = 0, y = 2$ and $y = 3 - x^2$. The density σ of D at point (x, y) is twice the distance from the point to the y-axis. Evaluate the mass of D.'

The standard definition of flux in the context provided in Figure 4 is that of an oriented surface integral. The surface is projected onto a coordinate plane and with the aid of a

Evaluate the upward flux of $\mathbf{F} = \langle 2, y, 2\rangle$ through S, where S is that portion of the surface $y = z - x^2$ which projects onto region R shown in the diagram.

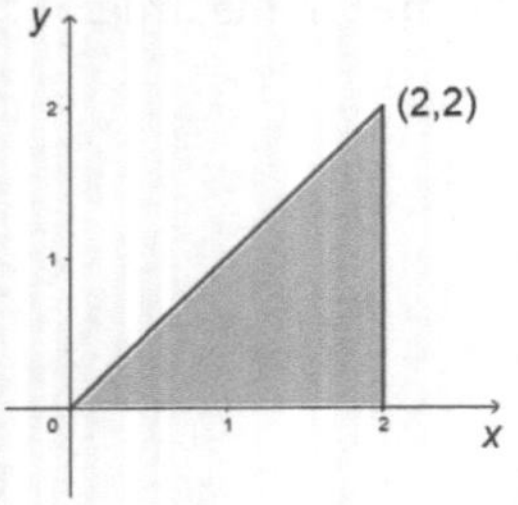

Figure 4. Decision-making and getting information from a diagram.

conversion factor it is reframed as a double integral and so solved. In the problem shown in Figure 4, the projected region is given which prompts the student to make decisions related to rewriting the surface equation as a function of x and y and how to proceed to set up the necessary double integral.

In our experience, students find oriented surface integral problems challenging even if presented in a familiar, textbook-standard format, such as 'D is the solid described by $\left\{(x, y, z) \in \mathbb{R}^3 : \sqrt{x^2 + y^2} \leq z \leq 1\right\}$. At the surface S of solid D the outward pointing normal vector is denoted by $\hat{\mathbf{N}}$. Given the vector field $\mathbf{F}(x, y, z) = (x - 1)\mathbf{i} + 2yx\mathbf{j} + xz\mathbf{k}$ calculate $\iint_S \boldsymbol{F} \cdot \hat{\mathbf{N}} dS$ directly, without using Gauss' Theorem.' Our example in Figure 4 does not represent a more challenging question than this standard item but rather one that is less susceptible to using an online calculator to solve (before having to make certain decisions) and is less likely to immediately match a standard solution template.

4. Takeaway

Our key takeaway from teaching and assessing vector calculus in 2020 and 2021 is that vector calculus is surprisingly accessible to open book (and open internet) assessment as long as the questions are designed with decision-making as the core guiding principle, backed up by using diagrams in useful ways and placing low emphasis on solving integrals. In 2020 and 2021, we allowed students to use online calculators and graphing software; however, the use of these tools has not yet been built into the curriculum more generally. Moving forward, we shall maintain open book testing and encourage the use of online tools.

Topics other than vector calculus could benefit from allowing options for decision-making in the assessment. An example from linear algebra is determining whether the representative matrix of a linear transformation is invertible or not. Options include finding the determinant by direct computation or by inference from other information, determining the kernel and/or image of the transformation, or geometrically describing how vectors are transformed. The assessment item could provide enough information for more than one of these options to be possible and the student is then placed in a position of having to both recognize the availability of options and decide how to choose among them.

5. Conclusion

All going well we shall not have to run a remote non-proctored test again with its associated weakness with regard to cheating. Similarly, we could go back to the traditional closed book test, but we shall not do so. Having been forced into the position of redesigning our assessment for the context of access to all notes and the tools available on the internet, we have been pleasantly surprised by the experience.

We designed assessment questions which placed decision-making, providing diagrams, providing information through diagrams and placing a low emphasis on computation at the core of the design process. This form of question in the open book, open internet context was well received by the students and the grade profile was not significantly different from previous years. The grading process was carried out in a standard manner although equity considerations remain, such as how to grade when one student solves an integral by hand and another uses Wolfram Alpha.

Certain types of questions we used to ask are no longer suitable for the open book, open internet environment, such as ones heavily reliant on computation or on drawing a diagram, in essence any question easily outsourced to an advanced calculator (see also Bailey et al., 2020). Proficiency at vector calculus is not demonstrated by computation, however (Bollen et al., 2015). We argue that it is demonstrated by information interpretation, recognizing available options and effective decision-making. There is not a great deal of research taking an overview on the teaching, learning and assessing of vector calculus (Martinez, 2019), although valuable work has been done on specific topics, such as line integrals (Mallet, 2013) or double integrals (Trigueros & Martínez-Planell, 2018). We hope that our experiences discussed here contribute to this valuable field.

In 2022 and beyond when we hope to be back in a physical classroom we shall retain the best features of our new assessment design. Our institution has a bank of Chromebooks available for assessment purposes. With forward planning, these can be set up to allow access to a virtual desktop and certain white-listed online resources so that students can use computer graphing packages, powerful online calculators and have access to certain course documents such as lecture slides. Planning has already started for the 2022 vector calculus test to be written in this form.

In March 2020, we greeted the new situation with dismay but swiftly realized that the difficult process of redesign was forcing us to improve our assessment offering. While we shall be relieved to put much of the 2020 and 2021 global experiences behind us we are pleased with the changes we have wrought in this aspect of our work and are excited to move forward.

Disclosure statement

No potential conflict of interest was reported by the author.

ORCID

Tracy S. Craig http://orcid.org/0000-0002-3653-2995

References

Bailey, T., Kinnear, G., Sangwin, C., & O'Hagan, S. (2020). Modifying closed-book exams for use as open-book exams. *OSF Preprint*. https://doi.org/10.31219/osf.io/pvzb7

Bollen, L., van Kampen, P., & De Cock, M. (2015). Students' difficulties with vector calculus in electrodynamics. *Physical Review Special Topics-Physics Education Research, 11*(2), 020129. https://doi.org/10.1103/PhysRevSTPER.11.020129

Caglayan, G. (2016). Teaching ideas and activities for classroom: Integrating technology into the pedagogy of integral calculus and the approximation of definite integrals. *International Journal of Mathematical Education in Science and Technology, 47*(8), 1261–1279. https://doi.org/10.1080/0020739X.2016.1176261

Campuzano, J. C., Roberts, A. P., Matthews, K. E., Wegener, M. J., Kenny, E. P., & McIntyre, T. J. (2019). Dynamic visualization of line integrals of vector fields: A didactic proposal. *International Journal of Mathematical Education in Science and Technology, 50*(6), 934–949. https://doi.org/10.1080/0020739X.2018.1510554

Craig, T. S. (2021). Open book vector calculus assessment: suggested design principles. In *Proceedings of SEFI 49th Annual Conference*, September 13–16, Berlin. https://sefi2021.eu/

Dimiceli, V. E., Lang, A. S., & Locke, L. (2010). Teaching calculus with Wolfram| alpha. *International Journal of Mathematical Education in Science and Technology, 41*(8), 1061–1071. https://doi.org/10.1080/0020739X.2010.493241

Iannone, P. (2020). Assessing mathematics at university: Covid-and beyond. *Newsletter of the London Mathematical Society, 490*, 34–40. https://www.lms.ac.uk/sites/lms.ac.uk/files/files/NLMS_490_for%20web2_1.pdf

Jungic, V. (2020). Making calculus relevant: Final exam in the time of COVID-19. *International Journal of Mathematical Education in Science and Technology*, 1–3. https://doi.org/10.1080/0020739X.2020.1775903

Khan, R. N. (2015). Assessments: An open and closed case. *International Journal of Mathematical Education in Science and Technology, 46*(7), 1061–1074. https://doi.org/10.1080/0020739X.2015.1076534

Mallet, D. G. (2013). An example of cognitive obstacles in advanced integration: The case of scalar line integrals. *International Journal of Mathematical Education in Science and Technology, 44*(1), 152–157. https://doi.org/10.1080/0020739X.2012.678897

Martinez, A. P. (2019). *Expert versus novice strategy Use during multiple integration problems* (Doctoral dissertation, Texas A&M University-Commerce. https://www.proquest.com/openview/652495d3ba472c514800e08c6180d9a2/1?pq-origsite=gscholar&cbl=18750&diss=y)

Pepper, R. E., Chasteen, S. V., Pollock, S. J., & Perkins, K. K. (2012). Observations on student difficulties with mathematics in upper-division electricity and magnetism. *Physical Review Special Topics-Physics Education Research, 8*(1), 010111. https://doi.org/10.1103/PhysRevSTPER.8.010111

Smith, G., Wood, L., Coupland, M., Stephenson, B., Crawford, K., & Ball, G. (1996). Constructing mathematical examinations to assess a range of knowledge and skills. *International Journal of Mathematical Education in Science and Technology, 27*(1), 65–77. https://doi.org/10.1080/0020739960270109

Trenholm, S. (2007). An investigation of assessment in fully asynchronous online math courses. *International Journal for Educational Integrity, 3*(2). https://doi.org/10.21913/IJEI.v3i2.165

Trigueros, M., & Martínez-Planell, R. (2018). *On students' understanding of Riemann sums of integrals of functions of two variables*. INDRUM 2018 proceedings :64-73.

OPEN ACCESS

No exam: assessment of third-year engineering students on the basis of self-generated statistics cases

Nelly Litvak and Fulya Kula

ABSTRACT
Through the global pandemic, the single greatest challenge at universities has been the move towards digital assessment. In this classroom note, we describe our new no-exam assessment setup in a Statistics course for third-year bachelor Mechanical Engineering students. The main idea is to assess the students based on self-generated problems. Crucially, we supported students by providing a very clear structure of the course material, learning objectives, and requirements. At the same time, we left sufficient space for the students to tune the final assignment to their interests, creativity, and mathematical skills. In our experience, this setup makes assessment meaningful and enjoyable for both the students and the teacher and does not need to demand excessive time investment on the teacher's side. We strongly believe that approaches like ours will have potential lasting effects on the diversity of assessment, quality of learning, and, last but not least, the appeal of Statistics for future engineers.

1. Introduction

Through the global pandemic, the move towards a digital assessment has become a paramount issue, with heated discussions on campus splashing out to the media. This issue is particularly prominent in undergraduate mathematics, where the assessment is predominantly based on a standard written test (Iannone, 2020). In this classroom note, we describe a new assessment setup of an introductory Statistics course for the class of 162 third-year undergraduate Mechanical Engineering students at the University of Twente. Instead of giving a traditional exam, the students were asked to create their own real-world Statistics cases.

Previous studies show that the assessment methods based on one-time measures, such as a traditional exam, are non-dynamic and fail to explore the developmental nature of learning (Chance & Garfield, 2002). Accordingly, the research has introduced assessment methods such as writing assignments (Woodard et al., 2020), activity-based statistics learning (Gnanadesikan et al., 1997), and projects (Garfield, 1994). One of the main focus points of the pre-pandemic research stands as the assessment of students' reasoning in Statistics (Garfield & Ben-Zvi, 2008) and calls recommended for this purpose implied the move towards alternative assessment techniques (Chance & Garfield, 2002). This invitation became emergent with the rapid and compulsory transition to online instruction throughout the pandemic. We contribute in this direction with our new assessment setup as a viable solution during the post-pandemic. Beyond providing an attractive digital assessment tool, such setup generates a resource of deeper insight into the students' understanding (Garfield & Ben-Zvi, 2008).

This classroom note is organized as follows. First, in Section 2 we describe the workflow of the course. The core of this article in Section 3, where we explain how we organized the assessment. In Section 4, we provide some examples and reflect on the results. Finally, in Section 4 we discuss the strengths and pitfalls of our approach and its possible lasting effects.

2. The course workflow

Presenting the course material. The course covered the basic material from the definition of probability to the single-sample hypothesis testing, with a workload of 2.5 ECTS, equivalent to 70 hours. We used book (Montgomery & Runger, 2018) and the complete series of short video lectures (Litvak, 2020) (openly available on YouTube). To trigger the students' interest, most examples in the course were specifically designed for the engineering profession. We have divided the material into five *Units*; the Units are listed in Table 1 together with examples of learning objectives that will be explained later.

The students studied the book and the videos independently, at a pace of about one Unit per week. For each Unit, we gave an *interactive (online) class*, where students answered multiple-choice questions, using the interactive platform *Wooclap*. We also held several online Q&A's for individual questions.

Structuring students' practice during the course. We ensured that the students kept practicing throughout the course by introducing *Quizzes* and *Problem Sets*. These activities, described in more detail below, were mandatory but did not affect the grade. The obligation was enabled through self-regulating prerequisites, facilitated automatically through the educational platform *Canvas*, and schematically depicted in Figure 1. For example, Quiz Unit 2 becomes accessible for a student only after Quiz Unit 1 is completed. At the end, the (fully online) Final Assignment becomes accessible only after completing all the prerequisites.

The *Quizzes* consisted of 4–6 standard problems, based on Mechanical Engineering examples, designed by us and not findable on the Internet. The Quizzes were implemented using *formula questions* in Canvas: the students had the same problems with different numerical values, so they could discuss together but the final answers were unique per quiz attempt. We allowed unlimited attempts and required a score of at least 60% to pass the Quiz.

For the *Problem Sets*, we used the design from Mazur (2018) that consists of the *individual phase* at home (the students solve a set of homework problems individually without discussing with others) and the *group phase* in class (the students discuss in small groups, mark their errors, and receive solutions from the teacher). It was mandatory to submit the solutions after the individual phase and to attend the group phase.

Table 1. The units and the examples of learning objectives with corresponding points.

Unit #	Unit title	Learning objective examples	points
1	How to quantify randomness? Probability, mean and variance	1.5 Apply the definition of independent events, and explain your computations	1
		1.12 Compute the variance and the standard deviation using a probability density function, and explain your computations	2
2	What is 6σ in 'lean 6σ'? Probability models	2.10 Reason in which real-life situations the Normal distribution is applied, and explain the meaning of its parameters	2
		2.11 Apply the Normal distribution, and explain your computations	2
3	What an engineer must know about the sample mean? – Linear functions and limit theorems	3.9 Apply the fact that a linear function of normal random variables has a normal distribution	2
		3.10 Recognize the situation where the Central Limit Theorem is applicable, and state when the normal approximation can be used based on this theorem	2
4	Plus-minus how much? Data summaries and confidence intervals	4.2 Compute sample variance and sample standard deviation	1
		4.16 Execute and explain computations for the confidence interval for the population proportion, using the normal distribution	2
5	What does it mean that 'the data is significant'? Hypothesis testing	5.2(b) One-sample t-test for the mean of the normal distribution: Formulate the null-hypothesis and the alternative	2
		5.2(d) One-sample t-test for the mean of the normal distribution: Accept or reject the null-hypothesis based on the appropriate critical region, or apply the notion of p-value	3

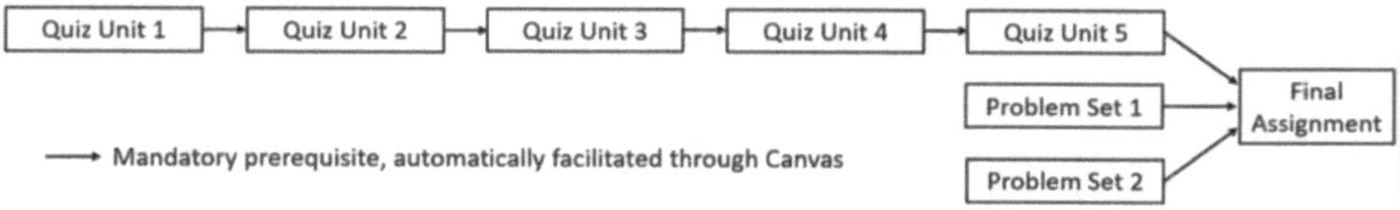

Figure 1. The workflow of the course. The Quizzes are graded automatically. For Problem Sets, only sub-mission and attendance to the group discussion is required.

3. The no-exam assessment setup

The core of this classroom note lies in the Final Assignment replacing the written test. We will first describe the Final Assignment, and then explain the requirements and the grading.

Final Assignment. We asked the students to do the following:

Find own cases from Mechanical Engineering, where the knowledge gained in the Statistics course was useful. The cases could be inspired by other courses, internships, research projects, etc., or other relevant real-world examples.

Formulate clear Statistics problems for the chosen cases. Crucially, we specified the volume and the coverage of the material by asking students to use a given number of the *points of learning objectives*, which we will explain in the next paragraph.

Create complete solutions to these problems by following three steps. We explain these three steps below.

Points of learning objectives. We have formulated very specific learning objectives, several examples are given in Table 1. Each learning objective refers to an exact topic and requires performing either very specific calculations (e.g. 2.11, 5.2(d)) or reasoning (e.g. 2.10, 3.10). One Unit contained about fifteen learning objectives.

To define the volume and the coverage of the Final Assignment, inspired by Faber and Dresscher (n.d.), we assigned 1–3 (usually, 2) *points* to each learning objective, see examples in Table 1. The points are approximately proportional to the corresponding weights we usually give in a standard written test. We required to use a minimum of 25 and a maximum of 30 points of learning objectives in the Final Assignment. In addition, we set minimal required points from each Unit, summing up to 21, and leaving the remaining 4–9 points for students' choice.

The students learned to relate problems to the learning objectives in Quizzes and Problem Sets. In Quizzes, after submission, the list of learning objectives for each problem appeared as a comment, and we explained to the students how to use this as a hint for the correct solution. In Problem Sets, during the discussions in the group phase, the students received a list of learning objectives for each problem and were encouraged to use this information in the discussion. They followed this advice because they wanted to practice for the Final Assignment.

Three steps of the solution. We asked the students to present their solutions in three steps, given in Table 2 below. These steps are inspired by Polya's four steps approach for solving mathematical problems (see e.g. Iannone, 2020). We adapted this approach to Engineering students by combining Polya's steps 1 and 2 and formulating the steps in a less abstract and more actionable way.

The students practiced the three steps in the Problem Sets. As in Mazur (2018), in the individual phase, the evidence of effort was that all three steps were worked out in all problems. As an example, the teacher presented the solutions using the three steps as well.

Grading using rubrics. The criteria that we used for grading are provided in Table 3. Importantly, we required that *each* criterion is met with *at least* 50% quality. For example, if Step 3 is missing in half of the problems, then the student gets a *zero* for this criterion. Moreover, the rule was that *failing one criterion resulted in failing the assignment.* This choice is motivated by our previous experience that the students in this course often aimed for a pass rather than a high grade. We wanted to encourage completeness of the work and to avoid that the students strategically skipped some parts to score just enough for a pass.

Table 2. The three steps of the solution.

Step 1	Explain in words how you would approach the problem: what needs to be computed? Which probability models are relevant and how? Which method will you apply (e.g. which statistical test) or which result will you apply (e.g. Central Limit Theorem)? Which steps will bring you to the answer?
Step 2	Solve the problem: Perform the computations as you described in Step 1.
Step 3	Look back at your solution and the answer: Is the answer reasonable/expected? Why or why not? If the answer does not seem correct, what could have gone wrong? What difficulties did you experience solving the problem? Which learning objectives are relevant for solving this problem?

Table 3. Grading criteria and their weights.

Criterion	%
The originality and the quality of the problem formulations	15%
The match between the problem formulations and the stated learning objectives	20%
The completeness and correctness of Step 1	20%
The completeness and correctness of Step 2	20%
The completeness and correctness of Step 3	10%
Video[a]	15%

[a]We have asked the students to upload a 4-minutes video where they explain their assignment. This is an unusual and somewhat controversial part of the assignment, and we will refer to this in Section 4.

To be consistent in the grading, the two teachers first graded two randomly chosen assignments together. Then, each student was graded by one of the teachers, and we have discussed questionable cases when needed.

4. Results

Quality of submissions. The cases were very interesting and varied, from material science and renewable energy to artificial intelligence and digital twins. Many assignments referred to recent engineering literature, see example in Figure 2 by student T. Edelijn. As customary in the Netherlands, we offered two attempts. After the first attempt, taken by 144 out of the 162 students, only 9 students failed to meet the passing grade of 55%; the average score was 75.57%, with a standard deviation 15.35%.

Importantly, the students did not make typical errors that we observed in the previous years when we used standard formative written tests! For example, the greatest majority of the students formulated the null – and the alternative hypothesis correctly and applied the correct test. In the previous years, these were sources of very common errors, even when the students used a formula sheet.

Experience with the learning objectives. The volume and the coverage of the students' assignments were as we intended it to be, that is, comparable to the standard written test in previous years. Since the points of learning objectives (see Table 1) were the only method we used to regulate the volume and the coverage, we conclude that this method was effective for this purpose.

> Fused deposition modelling (FDM) is a form of 3D printing which is getting more and more attention. Shape memory materials highly rely on the FDM process to achieve an accurate structure. The role of operational parameters as printing velocity and layer height is investigated by Barletta et al.[a]. They found that the layer height greatly influences the compression load the structure could endure. Therefore, accuracy is of the highest importance. The manufacturer of the machine claims that the accuracy of each layer is within 20 microns in 95% of the cases. Some students checked all individual layers of their 3D printed product and found: sample mean 212, sample variance 31, and n = 200. Is there enough statistical evidence to substantiate this claim?

[a] M. Barletta et al., 4D printing of shape memory polylactic acid (PLA) components: Investigating the role of the operational parameters in fused deposition modelling (FDM) , 2021.

Figure 2. An example of a problem generated by a student. The solution uses the χ^2 test for the variance of the normal distribution.

Another effect of the learning objectives was shifting the emphasis from computations to methodology. In written tests, in our experience, the students were very focused on calculations and did not pay attention to which technique they used and why. Now they were required to state the technique explicitly. This introduced an extra reflection step for the students and gave us the tool to assess the depth of their understanding. For example, a common error in the assignment was claiming learning objective 3.10 instead of 3.9 (see Table 1). This is a subtle error because calculations are the same in both cases. We used this to distinguish between excellent and average work, and could easily justify to the students that correct calculations did not necessarily result in a high grade.

Experience with the three steps of the solution (see Table 2). In most of the students' submissions, the solutions presented in three steps were clear, complete, and pleasant to read. We allowed some freedom of interpretation and did not reduce the score if, for example, information that belongs in Step 1 was included in Step 2.

We noticed that Step 3, the reflection on the answer, was hard for the students. Very commonly they wrote 'the answer makes sense' but their explanation did not reflect on the methodology. Also, there were some typical errors in Step 3, for example, interpretation of a confidence interval as if it was a prediction interval (we note that prediction intervals were not included in the course).

Overall, this format of the solution gave us one more tool to differentiate between excellent and average work by the quality of their arguments.

Experience with the rubrics. We found grading with rubrics as in Table 3 very suitable for this assignment because the resulting grades matched our general impression of the students' work. No student rendered their grade unfairly low and the work that received high grades was obviously of excellent quality.

In the criterion '*The originality and the quality of the problem formulation*' we considered 'originality' to be high when the content of the problems was based on own experience, and had little overlap with Quizzes and Problem Sets, while 'problem formulation' was required to be unambiguous and formally correct. We noticed that many students scored high on 'originality' and low on 'problem formulation', or the other way around. In the future, we will consider splitting these into two different criteria.

As discussed above, the criterion 'The match between the problem formulations and the stated learning objectives' was very informative for assessing a deeper understanding of statistical techniques. However, grading for this criterion was very time-consuming because the vast majority of the students identified at least some learning objectives incorrectly, which resulted in complicated calculations of their scores. In the future, we will keep this criterion but with a simplified version for scoring.

The *Video*, we believe, is potentially a powerful assessment tool, but it was of limited added value in this setup because most of the students simply summarized what already was written in the assignment and did not demonstrate any additional understanding in the video. Besides, some students strongly disliked the video. We will reconsider how to use the video, if at all, in the future.

Students' feedback. Based on the panel evaluation report, open comments in the standard course evaluations, and informal feedback from the students, they appreciated the assignment-based assessment but reported that it was too much work. For this reason, some of them said to have preferred a written test. We agree about reducing the workload in the future and can do so without damaging the comprehensiveness of the assessment.

We believe, however, that one reason for the perceived high workload was that the students learned better, which is evidenced, for example, by mostly correct application and explanation of methodology. Clearly, such deeper learning takes more time! We are very pleased that the assignment motivated the students to spend time on the course. Several students said that they enjoyed working on the assignment. We quote one of them: *I think that this form of examination is a neat way to make a student understand the content of the course without leading to high-stress levels and therefore found it a nice project to do.*

The students were very positive about the online activities of the course: the video lectures, the Quizzes, and the interactive classes.

In this classroom note, we presented qualitative results of our no-exam assessment setup. We close by discussing the strengths, weaknesses, and modifications of this approach, as well as its potential use in post-pandemic education.

Strengths, Weaknesses, Modifications. The strengths of this no-exam assessment setup are multidimensional, below we list those that we found most prominent.

- The emphasis was less on the recall of statistical procedures and more on exhibiting competence due to the clear focus on methodology rather than calculations, also accepted by the students.
- We strongly witnessed students' gains in terms of learning demonstrated by avoiding typical errors that occur in a written test.
- We observed students' positive attitudes towards Statistics in that many of them enjoyed creating problems close to their interests.
- Students' gradual work on the assignment, as well Quizzes and Problem Sets, created a spacing effect that is shown to produce better learning outcomes (Kornell, 2009).
- The students could plan their work and did not have the stress of a time-constrained test.
- The specification of learning objectives and the three steps of the solution were effective tools for assessing the depth of the students' understanding.
- Finally, the student-generated problems are a rich source for teachers to link Statistics with the engineering profession, invaluable for future courses.

As in every new initiative, there are points of improvement, some listed below.

- The students report that the workload of the Final Assignment was too high. A balance must be found between the workload and the comprehensiveness of the assessment.
- The grading of the students' Final Assignment took more time than the grading of a written test. The scoring systems must be simplified where possible.
- As usual with an essay/project type of assessment, plagiarism represents a potential limitation. We did not encounter plagiarism in our course, but this is a point of attention in the future.

Applications in post-pandemic education. We strongly believe that our no-exam setup is applicable in the future due to many benefits outlined above. Moreover, this setup is credible because it is been developed through years on the synergy of ideas and materials (e.g. (Mazur, 2018; Faber & Dresscher, n.d.)). The pandemic was merely catalysis for this work.

For the teachers who want to apply our no-exam assessment setup, we stress that the change in the assessment method cannot be considered separately from the ways of teaching. The assessment and the teaching described in this article are strongly aligned and connected. For example, even if the students cover only a fraction of the material in the Final Assignment, we ensure that they have practiced structurally throughout the course with all topics in the mandatory Quizzes and Problem Sets. Therefore, we would like to offer this entire setup as a road map. Some choices are specific to the teacher and the education programme, for example, the definition of learning objectives and the points assigned to them, as well as criteria and weights in the rubrics. Naturally, defining all the specifics for this type of assessment, together with a well-thought and designed course structure, requires a lot of work. However, once this setup is prepared, it can be used for many years.

Finally, in 2021 the course was given exclusively online. In the post-pandemic education, we will certainly hold the discussions in small groups (the group phase of the Problem Sets) on campus and would prefer to have interactive classes and Q&A's on campus as well. Nevertheless, the online tools such as video lectures, Quizzes, and digital platforms proved very useful, enabling, exciting, and greatly appreciated by the students. We will definitely continue to rely on these digital means in the future.

Acknowledgments

We thank Nico Rutten and Chantal Scholten for their great help in setting up the course. We are grateful to the Mechanical Engineering program, in particular Lisa Gommer and Genie Stoffels, for their support of our new assessment methods. We thank Erik Faber for sharing his ideas of assessment with us. Finally, we thank the students for their deliberate work on the course and their cooperation in exploring new assessment methods.

Disclosure statement

No potential conflict of interest was reported by the authors.

Funding

The author(s) reported there is no funding associated with the work featured in this article.

ORCID

Nelly Litvak https://orcid.org/0000-0002-6750-3484
Fulya Kula http://orcid.org/0000-0003-0367-1099

References

Chance, B. L., & Garfield, J. B. (2002). New approaches to gathering data on student learning for research in statistics education. *Statistics Education Research Journal* [Online], *1*(2), 38–41.

Faber, E. J., & Dresscher, D. (n.d.). *Introduction to physical systems.* https://osiris.utwente.nl/student/OnderwijsCatalogusSelect.do?selectie = cursus&cursus=202000968&collegejaar=2020&taal = en.

Garfield, J. B. (1994). Beyond testing and grading: Using assessment to improve student learning. *Journal of Statistics Education* [Online], *2*(1). https://doi.org/10.1080/10691898.1994.11910462

Garfield, J. B., & Ben-Zvi, D. (2008). Assessment in statistics education. In J. B. Garfield & D. Ben-Zvi (Eds.), *Developing students' statistical reasoning: Connecting re-search and teaching practice* (pp. 65–90). Springer.

Gnanadesikan, M., Scheaffer, R., Watkins, A., & Witmer, J. (1997). An activity-based statistics course. *Journal of Statistics Education*, *5*(2).

Iannone, P. (2020). Assessing Mathematics at University: Covid-and beyond. *Research Gate preprint*. https://www.researchgate.net/publication/342765751.

Kornell, N. (2009). Optimising learning using flashcards: Spacing is more effective than cramming. *Applied Cognitive Psychology: The Official Journal of the Society for Applied Research in Memory and Cognition*, *23*(9), 1297–1317. https://doi.org/10.1002/acp.1537

Litvak, N. (2020). Statistics: Why these formulas make sense? *YouTube* playlist. https://www.youtube.com/playlist?list=PLMc0PsvITOku19afGWLhtA63gWPZpWBuK.

Mazur, E. (2018). *Applied Physics 50A Syllabus*. Retrieved July 30, 2021, from https://canvas.harvard.edu/courses/29611.

Montgomery, D. C., & Runger, G. C. (2018). *Applied statistics and probability for engineers* (7th ed.). Wiley. ISBN 9781119585596.

Woodard, V., Lee, H., & Woodard, R. (2020). Writing assignments to assess statistical thinking. *Journal of Statistics Education*, *28*(1), 32–44. https://doi.org/10.1080/10691898.2019.1696257

Utilizing discussion boards for test questions: opportunities for students' mathematical creativity and uniqueness

Miloš Savić

ABSTRACT

Taking the challenges presented by shifting to online instruction mid-semester due to the Covid-19 global pandemic, I tried to utilize prompts on discussion boards as test questions. The prompts for these questions were to create a triple integral satisfying certain conditions, and when a person posted their integral, no one else could use the same function or bounds. What occurred was a dream to me as an educator – I got to see students being creative, collaborative, and thoughtful, while I enjoyed grading and seeing their uniqueness shine. In this article, I detail the integral prompt and another recent prompt for vectors that led to a merger between my teaching beliefs and actions. I also share about the multiple benefits for instructors, including seeing mathematically clever solutions, and students, including reducing test anxiety and reverse-engineering mathematics problems.

1. Introduction

For years, I have been advocating for incubation, or taking a break after a hard problem-solving session, in tertiary classrooms (Savić, 2015, 2016). This is part of my (and our research group's) devotion to fostering mathematical creativity in every classroom (El Turkey et al., 2018; Omar et al., 2019; Tang et al., 2020). A challenge to incubation is that it runs counter to timed tests. Although small amounts of stress may yield some creative ideas, tests are equivalent to anxiety-inducing events for most students (Boaler, 2014). I also walked around my class to keep students from cheating or copying from one another, which possibly contributed to that test anxiety. However, working together was one of my main goals for every class I teach, which runs counter to watching them for cheating. Why can't collaboration be possible while also emphasizing individuality? Finally, tests traditionally were about the answer, and trying to compute that answer as fast as one can (Stipek et al., 2001). This is the furthest possible from incubation and can lead to beliefs about oneself as not being fast (and hence 'math smart') and about the field as strictly focused on products and fast processes.

I had all these beliefs about tests swimming in my head for a long time, yet I kept introducing traditional tests and quizzes in my classes. In the spring semester of 2020,

COVID-19 shifted our instruction from in-class to online and lined up with my second test for the semester. It was for this test that I tried a technique that will continue for a long time – I posted a prompt on a discussion board (or forum) and laid out requirements for the students' responses, including that their responses had to be unique to each other. This caused a major shift in my approach to tests.

2. The first attempt at discussion-board problem posing

The first test question, from my multivariable calculus course, was as follows:

Create a triple integral in Cartesian coordinates (x, y, z) so that
1) there is a mixed term in the function $f(x, y, z)$ (i.e. you cannot have a function like $f(x, y, z) = x^2 + y^3$ because the x and y terms are separated),
2) that one of the bounds of the 3D region has to depend on the other two (i.e. you cannot have all bounds being numbers),
3) the integral equals 217, and
4) each of your integrals has to be unique to one another. Once an integral is posted, then that function and bounds cannot be used again.

Please post the integral on the discussion board using the 'Insert Math Equation' button, attach a file of your work proving that the integral equals 217, and drawing/demonstrating what your bounds are. Good luck and enjoy!

They were given 48 h to concoct a function and bounds, prove that their integral equalled 217 with all required work, and create a drawing that showed what their bounds may look like. Although most calculus courses offer only linear, straight-forward approaches to the content, the kind of reverse-engineering done in this test can be an important skill in mathematics and beyond (Beghetto, 2017). I had not seen it done anywhere in calculus, so I was not knowledgeable about what the results would be. What happened amazed me and the 34 students in the class as well.

Students approached the prompt with one of two different strategies. The *Factor* approach was to calculate a random function's integral and then multiply by a factor that made it 217 (in full disclosure, I would have done it this way, but it was not the most popular approach). The *Bound* approach was to leave one bound variable and calculate the final bound necessary to have 217. I saw students post their integral and work, and others could see what had been done, but I knew that even seeing the approaches is not enough to solve and create a different function. I could tell they also relied on each other for assistance, because when a post came in with the *Bound* approach, there were many others who used this same approach. I also promoted that they could submit all their scratchwork so that I could assign value to their process instead of only the final product. It was their final products that gave me a starting point to see what they were doing in their process and allowed for the grading to be more for feedback than for points.

It still feels odd to say this because of how time-consuming and frustrating grading has been for me in the past, but I really enjoyed grading the posts. I saw the multivariable calculus concepts that I was testing them on (including triple integration and three-dimensional boundaries), but I also saw how mathematically clever and creative students could be. Whether they were motivated by each other or by efficiency – some chose 0 as one of the bounds, while others chose the bound that depended on the other bounds in a way to cancel out some of the original function $f(x, y, z)$. There were still those that did not display the full knowledge of triple integrals, but they did approach the problem and tried their best to solve their own integral.

I assessed the problem using a 0–4 scale, where 0 was no attempt, 1 was any attempt, 2 was an attempt that show some understanding of the process, 3 was an attempt that had one of two parts correct (or almost correct): the calculation of the triple integral or the diagram of the bounds, and 4 was fully correct. I also graded some with a 3.5 since many aspects were approached correctly, but there was one major flaw in their problem to begin with. For example, a student had a bound from $z = 0$ to $z = 7$, but had a function $\frac{x^2y^2}{z^2}$ that they were integrating. I also gave ample feedback as to why they received the grade they did. I continued this discussion-board procedure, including many of the same requirements, in that same test with cylindrical and spherical coordinates and asked them to write a reflection on the processes of all three. This was the prompt:

> Write one page describing the ways you approached these last three Test 2 discussion questions. In particular, how did you think of your functions and bounds? What was common and what had to be different for each? How did you visualize the bounds?

Students wrote about picking bounds that would help them minimize error and be efficient, but also that the cylindrical and spherical coordinates were different than Cartesian because the angles are bounded. One student explicitly mentioned that they had difficulty and switched bounds, which is another piece of evidence that the longer time for finishing the question allowed for more freedom to learn from difficulties. All four parts were worth 4 points, and I also had a true/false exam in class for the final 4 points, thus totalling 20% of their final grade.

3. Benefits of the discussion-board problem posing

It was incredible to have students create and solve their own problems, especially for such a high-stakes situation. There were so many positives that I will continue the practice in every class. First, I got a chance to allow incubation in my classroom. Giving the students 48 h allowed them to further think about what and how they would approach the problem. In a more recent class, I extended the time to as much as seven days for them to attempt, reflect, and try as many times as they like. Second, the format allows them to use any resource that they would like. It is quite difficult to find a specific problem online that satisfies my requirements, so I can allow them to use anything, including one another's discussion posts, as a resource. Although there are restrictions – students cannot write or copy the post for one another – they are now learning from each other, so there is a sense of collaboration while allowing for individuality. I do realize that students could use computers or websites that can calculate these integrals, but I consider that a resource that the students could leverage and learn from. If I am interested in their process and keep reiterating that I appreciate their thinking and processes, I think that they will be more honest and try to do these processes on their own. This is echoed in Bretag et al. (2019), where the authors stated that students' cheating was 'primarily influenced by dissatisfaction with the teaching and learning environment,' (p. 1848) and offer to counter this with 'supporting students in learning to navigate… a highly connected and networked world, in which sharing and collaboration are an increasing part of professional practice' (p. 1850).

Third, it promotes a holistic view of problem solving. They are the ones that create and solve the problem, and the math education literature on problem posing has stated benefits, such as increased motivation, that are long-lasting (Hošpesová & Tichá, 2015; Silver, 1997). Many students see solving problems as a linear process, so showing 'behind the curtain' allows them to look at problems differently and perhaps dissect what is going on in other math problems. Fourth, it is *their* problem and solution, and I emphasize this because ownership and agency in mathematics can be crucial for developing students' mathematical identities. If they've created this problem, what else can they create mathematically?

There are also benefits for the instructor. The requirements that I posed are not too lengthy, so I had the benefit of saving time creating a test that I believe will show their knowledge and problem solving for the course. As I stated before, the grading was exciting because I got to see clever and creative mathematical moves, something that I do not normally get to see in a traditional test. I do want to say that the grading could get a bit lengthier at times because I was looking at a lot of the scratchwork to give pointed feedback. For example, I estimate that it took five hours on 34 students' first question. However, for every question I've assigned, I've gotten momentum that has allowed for quicker grading times, including seeing similar errors or successes. For larger class sizes, I would recommend finding a balance that is right for you of including these items with other types of assessments. I also liked that I was not hovering over them like in a traditional test, and the uniqueness requirement allows me to trust that they are creating something that is their own. Especially in a triple integral situation, a change in bound or function causes a lot to change in the integration itself.

4. The following semester and beyond

With the same approach, I taught calculus of vector-valued functions, parametric, polar, and sequences/series the following semester. One of the prompts of my first test in that class was the following:

Given the vector $3, \sqrt{5}, 1$, create another unique (to your peers) vector (with non-zero components) so that the dot product of the two vectors is 10. Then, find the:

1. Cross of the two vectors
2. Unit vectors of both
3. Angle between the two vectors
4. Projection vector of your vector onto the given vector.
5. Area of the triangle created by the vectors and the subtraction of the vectors.

You do not have to simplify certain things (keep square roots and all the mess that is inside) if some of your calculations get complicated. Finally, I would like you to draw all of these vectors on one 3D axis. Good luck!

After grading the test question, I noticed that everyone had the same vector for their projection vector, even though they had different vectors to satisfy the rest of the prompt. This was parlayed into a group quiz question, where I asked them why this phenomenon occurred. It reinforced even more the underlying importance of the dot product as not just a simple equation but one that can show the information of one vector on another, which is key for the subsequent major theorems towards the end of calculus (e.g. Green's and Stokes's Theorem). This uniqueness discussion-board prompt allowed me to have deeper conversations about key ideas. It can also be used as a reflective tool to demonstrate the theory – students get to see 30 examples of a concept and can think about the main concept instead of the minutiae or calculations.

Students have also praised the approach. It reduces the amount of anxiety caused by high-stakes testing, which can take up a large amount of their working memory. It also allows students who may not have understood either the prompt or the concept to get more understanding through their peers. Since there is uniqueness built-in, they still get a chance to do their own calculations and reflections. They have also expressed pride over what they created. Generating self-efficacy is difficult, so to have students feel like they can control, approach, and create mathematics can be a turning point in their STEM careers (Regier & Savić, 2020).

This approach is not unique to discussion boards or technology. One can set up these prompts and allow students to work with each other in-class or outside. There are advantages to technology that include the uniqueness of students' responses, but there could be other websites beyond the traditional discussion board such as Discord, GroupMe, or Flipgrid. However, I do enjoy that both the prompt and grading are seamless with Canvas (our learning management system). There are also many ways that students are communicating outside of class, so there is a chance that a prompt of uniqueness will be checked by others outside of class.

This pandemic has forced my teaching to push outside-the-box and to truly commit to what I've dedicated my life to researching – creativity. I saw more creativity in my students, got to create opportunities for them to think deeply, and celebrated their ownership of the material. Creating discussion boards that showcase students' uniqueness while demonstrating understanding of content will be in my pedagogical toolbox for the rest of my career.

Acknowledgments

I thank Casey Haskins for her support and comments on the first draft of this manuscript. I also thank my family and friends for their continued support.

Disclosure statement

No potential conflict of interest was reported by the author.

ORCID

Miloš Savić http://orcid.org/0000-0002-7123-9946

References

Savić, M. (2015). The incubation effect: How mathematicians recover from proving impasses. *The Journal of Mathematical Behavior*, *39*, 67–78.

Savić, M. (2016). Mathematical problem-solving via Wallas' four stages of creativity: Implications for the undergraduate classroom. *The Mathematics Enthusiast*, *13*(3), 255–278.

Beghetto, R. A. (2017). Lesson unplanning: Toward transforming routine tasks into non-routine problems. *ZDM*, *49*(7), 987–993. https://doi.org/10.1007/s11858-017-0885-1

Boaler, J. (2014). Research suggests that timed tests cause math anxiety. *Teaching Children Mathematics*, *20*(8), 469–474. https://doi.org/10.5951/teacchilmath.20.8.0469

Bretag, T., Harper, R., Burton, M., Ellis, C., Newton, P., Rozenberg, P., Saddiqui, S., & van Haeringen, K. (2019). Contract cheating: A survey of Australian university students. *Studies in Higher Education*, *44*(11), 1837–1856. https://doi.org/10.1080/03075079.2018.1462788

El Turkey, H., Tang, G., Savić, M., Karakok, G., Cilli-Turner, E., & Plaxco, D. (2018). The creativity-in-progress rubric on proving: Two teaching implementations and students' reported usage. *Primus*, *28*(1), 57–79. https://doi.org/10.1080/10511970.2017.1346735

Hošpesová, A., & Tichá, M. (2015). Problem posing in primary school teacher training. In F. M. Singer, N. F. Ellerton & J. Cai (Eds.), *Mathematical problem posing* (pp. 433–447). Springer.

Omar, M., Karakok, G., Savić, M., Turkey, H. E., & Tang, G. (2019). I felt like a mathematician: Problems and assessment to promote creative effort. *Primus*, *29*(1), 82–102. https://doi.org/10.1080/10511970.2018.1475435

Regier, P., & Savić, M. (2020). How teaching to foster mathematical creativity may impact student self-efficacy for proving. *The Journal of Mathematical Behavior*, *57*, 100720. https://doi.org/10.1016/j.jmathb.2019.100720

Silver, E. A. (1997). Fostering creativity through instruction rich in mathematical problem solving and problem posing. *Zentralblatt Für Didaktik Der Mathematik*, *29*(3), 75–80. https://doi.org/10.1007/s11858-997-0003-x

Stipek, D. J., Givvin, K. B., Salmon, J. M., & MacGyvers, V. L. (2001). Teachers' beliefs and practices related to mathematics instruction. *Teaching and Teacher Education*, *17*(2), 213–226. https://doi.org/10.1016/S0742-051X(00)00052-4

Tang, G., Savić, M., El Turkey, H., Cilli-Turner, E., Karakok, G., & Regier, P. (2020). Shifting pedagogical beliefs into action through teaching for mathematical creativity. In S. Karunakaran, Z. Reed, & A. Higgins (Eds.), *Proceedings of the 23rd Annual Conference on Research in Undergraduate Mathematics Education*, 959–965. http://sigmaa.maa.org/rume/RUME23.pdf

Mathematics assessment integrity during lockdown: experiences in running online un-invigilated exams

Steven Richardson

ABSTRACT

During the 2020 COVID-19 lockdowns, one of the primary challenges faced by mathematics educators was maintaining assessment integrity when replacing invigilated assessments with online assessments. These assessments presented students with the opportunity to engage in misconduct in a manner that ordinarily would not exist. This was particularly true when online proctoring tools were not adopted. In this article, I present observations from running online un-invigilated assessments in two first-year calculus-based classes at Edith Cowan University (ECU). I outline the key principles applied in structuring the assessments and identify four styles of question that stood out as being less impacted by misconduct behaviours. I then identify some common indicators of potential misconduct, describe the approaches used to investigate potential breaches, reflect on student responses during misconduct interviews, and highlight an important consideration when establishing panels to review appeals against misconduct decisions in mathematics.

1. Introduction

The closure of university campuses in response to the COVID-19 pandemic in 2020 generated significant challenges for academics who were required to transition to an entirely online mode of delivery and assessment within a short time period. One of those challenges was the need to replace large scale face-to-face invigilated assessments with online alternatives, without significantly compromising assessment integrity.

Face-to-face examinations have long been a cornerstone of university mathematics assessment (Iannone & Simpson, 2011; Thoma & Nardi, 2016). This is not just a consequence of conservatism, but the belief that these assessments provide an appropriate balance of efficiency, validity and fairness (Iannone & Simpson, 2012). The difficulty in monitoring students taking large scale assessments in an online setting creates a significant challenge in maintaining assessment integrity.

Butler-Henderson and Crawford (2020) provided a review of the literature related to online examinations and identified key research themes relating to perception, performance, anxiety, cheating, authentication and security, and technology. They found that the most prevalent focus of research into online examinations relates to cheating, with the majority focused on technical challenges rather than social aspects of cheating.

Online proctoring is an approach which offers the potential to enhance assessment integrity by monitoring student activity while they complete assessments. Although results have been mixed, there is growing evidence of the effectiveness of online proctoring. Dendir and Stockton Maxwell (2000) introduced online proctoring half-way through a teaching period in two courses and observed a significant drop in performance in every exam sat under proctored conditions. They also observed that Grade Point Average, as a measure of ability, was much more correlated with performance in the proctored exams than exams that were not proctored. Howard (2019) compared the performance of mathematics students sitting exams under different conditions, observing that although there was no significant difference in mean scores, students taking exams without proctoring took longer to complete their exams than those that were proctored. This was interpreted as evidence that, in the absence of proctoring, student had been spending time searching for answers/information online.

Although online proctoring has been utilized at several universities, this approach is not universally embraced due to the relatively high cost of these services, and concerns for students' privacy and potential anxiety over being constantly monitored (ul Haq et al., 2015). In the absence of proctoring there is no direct way to ensure that submitted work is that of the submitting student. Students have the potential to seek assistance from friends, family, tutors, online 'tutoring' sites such as Chegg, and online calculators such as Wolfram Alpha that not only present the solution to a problem, but the working as well. Nevertheless, it is important to consider approaches that could go some way toward reducing the utility of these resources and/or detecting instances where students have breached the assessment rules.

The ongoing exploration of online alternatives to invigilated assessment will continue to be relevant in years to come. At present, entirely online assessment continues to be necessitated by the ongoing COVID-19 pandemic. Closed borders and reduced availability of remote examination centres mean that students who ordinarily would be assessed face-to-face need to be assessed online. Beyond the pandemic there will be a continued push in many universities to move away from invigilated examinations. This is driven both by a desire to move toward authentic styles of assessment (Villarroel et al., 2018), as well as the desire to safeguard against a future COVID-19 type disruption to face-to-face learning and assessment. Academic misconduct research specific to mathematics will be necessary given a distinct lack of such research to date (Seaton, 2019).

The remainder of this classroom note is organized as follows. In Section 2, I outline the main features of the assessments that I used to replace face-to-face invigilated exams in two first-year calculus-based classes at ECU. In Section 3, I present four question styles that stood out as being less impacted by misconduct behaviours. In Section 4, I describe some common indicators of potential misconduct, describe the approach used to investigate potential breaches, and comment on the lack of understanding of mathematical processes that was encountered during the student appeals process. Finally, in Section 5 I highlight the key takeaways.

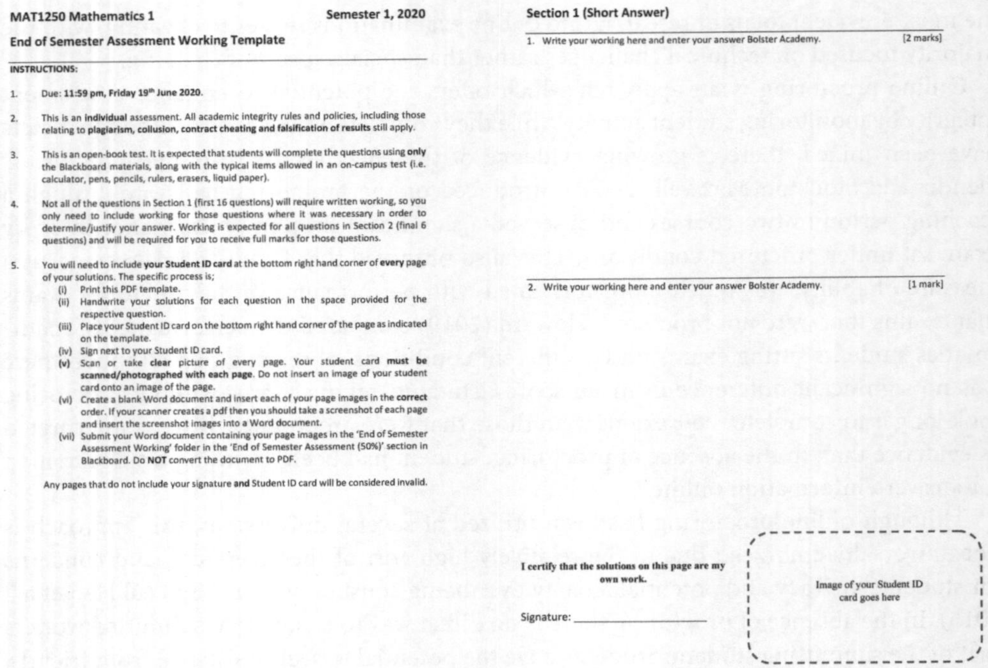

MAT1250 Mathematics 1 **Semester 1, 2020**

End of Semester Assessment Working Template

INSTRUCTIONS:

1. Due: **11:59 pm, Friday 19th June 2020.**

2. This is an **individual** assessment. All academic integrity rules and policies, including those relating to **plagiarism, collusion, contract cheating and falsification of results** still apply.

3. This is an open-book test. It is expected that students will complete the questions using **only** the Blackboard materials, along with the typical items allowed in an on-campus test (i.e. calculator, pens, pencils, rulers, erasers, liquid paper).

4. Not all of the questions in Section 1 (first 16 questions) will require written working, so you only need to include working for those questions where it was necessary in order to determine/justify your answer. Working is expected for all questions in Section 2 (final 6 questions) and will be required for you to receive full marks for those questions.

5. You will need to include **your Student ID card** at the bottom right hand corner of **every** page of your solutions. The specific process is;
   (i) Print this PDF template.
   (ii) Handwrite your solutions for each question in the space provided for the respective question.
   (iii) Place your Student ID card on the bottom right hand corner of the page as indicated on the template.
   (iv) Sign next to your Student ID card.
   (v) Scan or take **clear** picture of every page. Your student card **must be scanned/photographed with each page**. Do not insert an image of your student card onto an image of the page.
   (vi) Create a blank Word document and insert each of your page images in the **correct** order. If your scanner creates a pdf then you should take a screenshot of each page and insert the screenshot images into a Word document.
   (vii) Submit your Word document containing your page images in the 'End of Semester Assessment Working' folder in the 'End of Semester Assessment (50%)' section in Blackboard. Do **NOT** convert the document to PDF.

Any pages that do not include your signature **and** Student ID card will be considered invalid.

Section 1 (Short Answer)

1. Write your working here and enter your answer Bolster Academy. [2 marks]

2. Write your working here and enter your answer Bolster Academy. [1 mark]

I certify that the solutions on this page are my own work.

Signature: ____________

Image of your Student ID card goes here

Figure 1. The first two pages of a template for students to write their working.

2. Assessment design

My main objectives in designing an exam-replacement assessment were to examine content in a way that:

- would not be trivialized by the availability of online calculators (e.g. Wolfram Alpha) and other software tools (e.g. Matlab);
- would provide mechanisms to detect potential misconduct; and
- would provide a level of identity verification.

The assessment was conducted in two parts: (1) A traditional concept/calculation-based component; and (2) A short written component that was mandated for all classes in the School of Science at ECU.

The concept/calculation component was administered through the online assessment platform Bolster Academy (https://bolster.academy/) which was already being used for students to complete assignments and practice exercises. A week prior to the assessment students were provided with a PDF template that included instructions and space for them to write out working as they completed the assessment. The first two pages of the template are shown in Figure 1.

The Bolster Academy test was timed for three hours. During this time students were permitted to access materials on the unit Blackboard site (the LMS) but not any other materials. After the test students were given 12 h to photograph or scan each page of their working with their student ID placed in the bottom right hand corner, compile the images into a single document, and submit via the Blackboard site. The motivations behind this assessment design were as follows.

- Bolster Academy can randomize parameter values in the assessment questions. This meant that each student would have slightly different questions, making collusion more difficult.
- Having students submit their working allowed marks to be awarded when a final answer was incorrect. It also provided a means of identifying anomalies pointing toward potential misconduct. The responses in a student's working needed to match the responses entered into Bolster Academy to ensure that answers had not been changed after the end of the assessment.
- Including student ID in the photograph/scan of the working meant that it would not be possible for the working to have been completed by someone who did not have the student's ID card.
- Restricting the 'open book' nature of the assessment exclusively to unit materials meant that if students deviated significantly from the methods demonstrated in class then it could be expected that they would be able to explain their method.

There were ways in which students undermined some of the misconduct detection mechanisms by not following instructions as intended. Specifically:

- Some students used the 12-hour working submission period to re-write their working neatly. This meant that the working was not an accurate reflection of the sketches and scribbles which usually accompany the solution of mathematical problems. The working submission period was reduced to 30 min in subsequent iterations to discourage this behaviour.
- Rather than photographing their student ID with their working, some students pasted images of their student ID onto images of their working. This undermined the purpose of including the student ID, so these students were automatically interviewed in a video call after the assessment to check their understanding of the working they submitted.

The written component of the assessment was administered using the software Cadmus (https://cadmus.io/). This component was open for 24 h, and contained two questions:

(1) The solution to a problem was presented that contained deliberate errors. Students needed to identify those errors and indicate how they could be corrected.
(2) In 150–200 words students were asked to identify the topic/problem in the unit that they were most proud to have mastered, the module in which it was studied, and briefly describe the topic/problem and the main challenge(s) in mastering it.

The idea behind the second question was that a reflection response might provide a means of identifying students who had sought outside help from someone who did not know the specific structure of the class content. This didn't work well, as many students failed to address the question. It was unclear whether a poor answer was the result of misconduct, or a misunderstanding. For example, some students described a topic but did not

E.g. Indicate which of the following differential equations are both linear and separable.

E.g. Suppose that $z_1 = 8 - 2i$. Determine a number z_2 such that $z_1 z_2$ is real.

Figure 2. Two examples of conceptual questions used in the online assessments in semester 1 2020.

talk about their own challenges, while others wrote about personal challenges rather than mathematical challenges.

3. Successful question types

Across both assessment components, four question types stood out as being effective in assessing mathematical concepts in ways that would have been difficult to answer directly using online calculators and software tools.

The first two types were conceptual questions and worded questions. Conceptual questions can typically be answered without any (or very limited) working. Examples are shown in Figure 2. Worded questions require students to establish equations from the information provided. An example is shown in Figure 3. These question types are typically found to be more difficult than direct calculations by mid-range and weaker students, and results were similar in the online assessment. Internal statistical analysis demonstrated that the proportion of students who answered conceptual questions correctly was very similar to that observed in standard invigilated assessments, as was the working presented for worded questions.

The third question type required students to fill in missing terms and steps in a solution procedure. An example is shown in Figure 4. This was not a question style that I had used previously, however, I found it effective in the online context. Even if a calculator provides working to the problem, unless the steps are laid out identically to those in the question, it would still be necessary for the student to be able to match up the two solutions in order to fill in missing parts. Even if they were able to do this it would take much longer than working through the question as intended. Results for these questions were comparable to those observed in invigilated assessments where students were required to perform the corresponding calculation.

The fourth question type required students to identify errors in the working to a problem. Part of an example is shown in Figure 5. The ability to identify errors in working is not a skill that I had tested before, however, I found the question style to be discerning of different ability levels, ranging from students who stated that there were no errors, through to students who found five errors (out of a total of five deliberate errors). Despite having 24 h, and access to all unit materials, the average number of errors found was only 2.5. These questions are also not easily answered using online calculators. The only issue with this question type was that some students did not understand the 'follow-through' principle, and instead identified the first error, and then provided their version of the correct working from that point on (which in all such cases was still incorrect). Although they were provided with another example prior to the assessment, more opportunities to practice would likely address this problem in future.

A factory is located on the bank of a straigth river that is $w = 5$ km wide. There is a power station on the opposite bank $L = 8$ km upstream, which supplies the factory with electricity via the cable shown in the figure below.

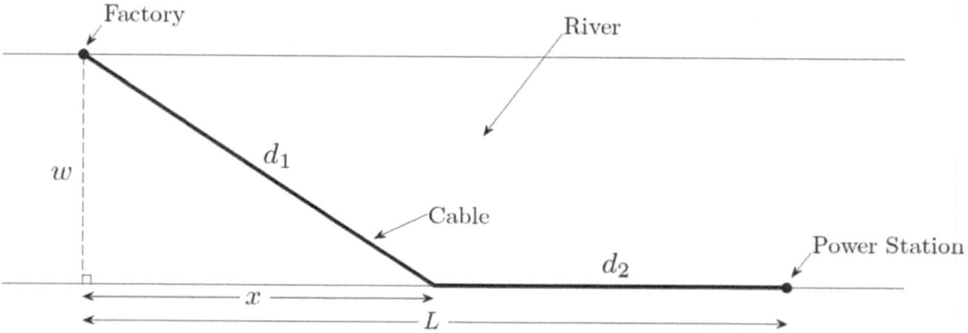

Above ground cable costs $\$20000$ per km to install, while underwater cable costs $\$60000$ per km to install. Given that the total cost of installing the cable in units of ten thousand dollars is given by

$$C = 6\,d_1 + 2\,d_2$$

determine the distance x in the above figure that will minimise the cost of the cable instalation. State your answer exactly.

Figure 3. A worded optimization question used in one of the online assessments in semester 1 2020.

4. Misconduct investigations

Despite the success of the question types outlined in Section 3, there were significant misconduct issues related to calculation questions. Seaton (2019) identified 6 features of mathematical responses that may indicate misconduct: strange working, correct answer; unusual wrong answer; unusual/unexpected right answer; inconsistent standard of answers; the right answer to a different question; and an answer from another level. The key indicators of potential misconduct observed in our assessments are listed below, with their alignment to those of Seaton (2019) indicated in parentheses.

- Skipping steps in working (strange working, correct answer). Figure 6 is a transcription of a student's response to an assessment question. The jump in reasoning from line (2)–(3) is significant, and not what I would expect.
- Using methods that were different to those used in class and distinct from solutions previously seen in invigilated settings (unusual/unexpected right answer; answer from another level). Figure 7 shows a solution that involves reasoning that I have never seen from a student at this level. For example, rewriting 2 as $\log_4(4^2)$ when going from line (6)–(7).
- Nonsensical steps that arrive at the correct answer (unusual/unexpected right answer). An example is given in Figure 8. Most likely the student obtained the answer from an online calculator but could not reconstruct the working.

Fill in the gaps (missing parts) in the working below.

Consider the initial value problem

$$\frac{dy}{dx} + 6\,y = -8\,e^{-4\,x}, \quad y(0) = 1$$

The differential equation is [▾] (i.e. of the form $\dfrac{dy}{dx} + P(x)\,y = Q(x)$) where $P(x) =$ ___ and $Q(x) =$ ___ . The integrating factor is given by

$$I(x) = e^{\int 6\,dx}$$

$$= \_\_$$

The integral $\int I(x)\,Q(x)\,dx$ evaluates to

$$\int I(x)\,Q(x)\,dx = \int \left(e^{6\,x}\right)\left(-8\,e^{-4\,x}\right)$$

$$= -8\int \_\_ \; dx$$

$$= \_\_$$

Hence the general solution to the differential equation is

$$y(x) = \frac{1}{I(x)}\left(\int I(x)\,Q(x)\,dx + c\right)$$

$$= \frac{1}{e^{6\,x}}\left(\_\_ \quad +c\right)$$

$$= \_\_ + \frac{c}{e^{6\,x}}$$

Imposing the initial condition $y(0) = 1$ we have

$$1 = -4 + c$$

$$\Rightarrow c = \_\_$$

Hence the particular solution to the initial value problem is

$$y(x) = \_\_$$

Figure 4. A fill-in-the-blank question used in one of the online assessments in semester 1 2020.

Students who accessed unauthorized material/calculators/assistance but submitted working that followed an expected procedure could not be detected.

The investigation of potential misconduct cases required each student to attend an online interview in which they were asked questions about their working. Unfortunately, there was a one month time delay between the assessment and the interview due to the time taken to complete the marking, report and process misconduct cases, and schedule

Consider the question and solution provided below. Identify any mathematical errors in the solution, and briefly state how the error would be corrected. You need to apply the follow-through principle, in that once you identify an error you carry on following the working as if the incorrect line of reasoning is correct.

Question: Solve the initial value problem

$$y'' - 4y' + 20y = 10x^2 + 4, \quad y(0) = 2, \; y'(0) = 0$$

Solution: The solution to the complementary equation

$$y'' - 4y' + 20y = 0$$

is obtained by first solving the characteristic equation

$$\begin{aligned} \lambda^2 - 4\lambda + 20 &= 0 \\ \Rightarrow \quad \lambda &= \frac{-(-4) \pm \sqrt{(-4)^2 - 4(1)(20)}}{2(1)} \\ &= \frac{4 \pm \sqrt{-64}}{2} \\ &= 2 \pm 8i \end{aligned}$$

Hence the general complementary solution is given by

$$y_c(x) = c_1 e^{8x} \cos(2x) + c_2 e^{8x} \sin(2x) \qquad (1)$$

Figure 5. Part of the working presented to students from which they were asked to identify errors.

Solve the following equation for x

$$4^{4x+2} = 9^{2x+3}$$

Express your answer exactly.

Student Response:

$$\begin{aligned} \ln\left(4^{4x+2}\right) &= \ln\left(9^{2x+3}\right) && (1) \\ (4x+2)\ln(4) &= (2x+3)\ln(9) && (2) \\ x &= \frac{3\ln(9) - 2\ln(4)}{4\ln(4) - 2\ln(9)} && (3) \end{aligned}$$

Figure 6. An example of missing steps in working between line (2) and line (3).

interviews. As a result, cases in which students demonstrated even marginal understanding of their working were dismissed, accounting for approximately 25% of cases investigated. For those who were unable to demonstrate any understanding of their working, the typical excuses included:

- They had forgotten everything since the assessment.
- They had written their working on other sheets of paper and copied it later to the submitted template. They did not copy all steps and had lost the original working.

Solve the following equation for x

$$4^{4x+2} = 9^{2x+3}$$

Express your answer exactly.

Student Response:

$$\begin{aligned}
\log_4\left(4^{4x+2}\right) &= \log_4\left(9^{2x+3}\right) && (1)\\
4x+2 &= (2x+3)\log_4(9) && (2)\\
4x+2 &= 2\log_4(9)x + 3\log_4(9) && (3)\\
2-3\log_4(9) &= 2\log_4(9)x - 4x && (4)\\
2-3\log_4(9) &= (2\log_4(9)-4)x && (5)\\
x &= \frac{2-3\log_4(9)}{2\log_4(9)-4} && (6)\\
&= \frac{\log_4(4^2)-\log_4(9^3)}{\log_4(9^2)-\log_4(4^4)} && (7)\\
&= \frac{\log_4\left(\frac{4^2}{9^3}\right)}{\log_4\left(\frac{9^2}{4^4}\right)} && (8)\\
&= \log_{\frac{81}{256}}\left(\frac{16}{729}\right) && (9)
\end{aligned}$$

Figure 7. An example of unusually sophisticated working, particularly from line (6) to line (7).

Solve the following equation for x

$$4^{4x+2} = 9^{2x+3}$$

Express your answer exactly.

Student Response:

$$\begin{aligned}
4^{4x+2} &= 9^{2x+3} && (1)\\
(4x+2)\log(4) &= (2x+3)\log(9) && (2)\\
\log(4x)2\log(4) &= 3\log(9)\log(2x) && (3)\\
4\log(x)2\log(4) &= 3\log(9)2\log(x) && (4)\\
4\log(4)x + 2\log(4) &= 2\log(9)+3\log(9) && (5)\\
4\log(4)x &= 4\log(3)+6\log(3)-4\log(2) && (6)\\
x &= \frac{6\log(3)-4\log(2)}{8\ln(2)-4\ln(3)} && (7)
\end{aligned}$$

Figure 8. An example of nonsensical working. Every step, with the exception of line (1) to line (2) and line (5) to line (6), is incorrect.

- They did not understand the method demonstrated in class and had learned a different approach which they had used in the assessment but had forgotten since.
- They had used additional notes during the assessment (which was not allowed).

We considered that, had they been able to solve the problem in the assessment, it was unlikely that they would not be able to make any meaningful comment about their working.

Several students appealed the misconduct finding against them through the student appeals committee. It became evident through this process that academics from non-numerate disciplines do not necessarily understand the role of working in mathematics. One student who submitted a solution similar to that in Figure 6 was exonerated based on the finding that 'whilst your answer did not show as many steps as a model answer would have shown, you did provide some working out in your answer to this question'. The committee did not understand that working serves a purpose beyond communication, as a means of keeping track of reasoning while solving a problem. They further did not appreciate how difficult and unlikely it would be to solve the problem without including the missing steps.

5. Conclusions and takeaways

My experience in conducting online assessment has highlighted the value of invigilated assessment in mathematics classes. Although it is possible to assess content in ways that are less trivialized by online calculators, without using online proctoring services it is not possible to efficiently ensure that a student did not access unauthorized assistance, particularly with large cohorts.

The main takeaways from my experience are:

- In the absence of online proctoring, a follow-up interview asking students to explain the thought processes underpinning their working is an effective means of authenticating a student's submission. The time between the assessment and the interview should be minimized. Administering this process would be inefficient for large cohorts.
- In the absence of follow-up interviews for all students, potential misconduct cases need to be identified and processed quickly to minimize the time between the assessment and a follow-up interview in which a student would be required to explain their working.
- Questions requiring students to fill-in-the-blanks or identify errors in working assess skills that are not broadly assessed in first-year mathematics classes (i.e. the ability to work through someone else's working). They are also less trivialized by the availability of online calculators, so are a good supplement to other question types in online assessments.
- Instructions need to be clear so that students do not inadvertently undermine integrity measures, and to provide clarity in misconduct investigations.
- Student appeals committees must include staff from a discipline cognate to that of the alleged misconduct. Staff from unrelated disciplines may not understand the context, which can result in decisions being based on incorrect reasoning.

Disclosure statement

No potential conflict of interest was reported by the author.

References

Butler-Henderson, K., & Crawford, J. (2020). A systematic review of online examinations: A pedagogical innovation for scalable authentication and integrity. *Computers and Education, 159,* 104024. https://doi.org/10.1016/j.compedu.2020.104024

Dendir, S., & Stockton Maxwell, R. (2020). Cheating in online courses: Evidence from online proctoring. *Computers in Human Behaviour Reports, 2,* https://doi.org/10.1016/j.chbr.2020.100033

Howard, D. (2019). Online testing integrity in a general education math course: A correlational study. *Dissertation Abstracts International Section A: Humanities and Social Sciences, 81*(7-A), 76–88.

Iannone, P., & Simpson, A. (2011). The summative assessment diet: How we assess in mathematics degrees. *Teaching Mathematics and its Applications, 30*(4), 186–196. https://doi.org/10.1093/teamat/hrr017

Iannone, P., & Simpson, A. (2012). A survey of current assessment practices. In P. Iannone & A. Simpson (Eds.), *Mapping university mathematics assessments practices* (pp. 3–16). University of East Anglia.

Seaton, K. A. (2019). Laying groundwork for an understanding of academic integrity in mathematics tasks. *International Journal of Mathematical Education in Science and Technology, 50*(7), 1063–1072. https://doi.org/10.1080/0020739X.2019.1640399

Thoma, A., & Nardi, E. (2016). A commognitive analysis of closed-book examination tasks and lecturers' perspectives. In E. Nardi, C. Winslow, & T. Hausberger (Eds.), *Proceedings of INDRUM 2016: First conference of the international network for didactic research in university mathematics* (pp. 411–420). University of Montpellier and INDRUM.

ul Haq, A., Jamal, A., Butt, U., & Majeed, A. (2015). Understanding privacy concerns in online courses: A case study of Proctortrack. In H. Jahankhani, A. Carlile, B. Akhgar, A. Taal, A. Hessami, & A. Hosseinian-Far (Eds.), *Global Security, Safety and Sustainability: Tomorrow's Challenges of Cyber Security. ICGS3 2015. Communications in computer and Information Science, 534* (pp. 137–150). Springer. https://doi.org/10.1007/978-3-319-23276-8_12

Villarroel, V., Bloxham, S., Bruna, D., Bruna, C., & Herrera-Sed, C. (2018). Authentic assessment: Creating a blueprint for course design. *Assessment and Evaluation in Higher Education, 43*(5), 840–854. https://doi.org/10.1080/02602938.2017.1412396

Mathematics and academic integrity: institutional support at a Canadian college

Josh Seeland, Lynn Ciplef, Caitlin Munn and Craig Dedrick

ABSTRACT

Academic integrity at our small Canadian college is informed by several key frameworks and centred around teaching, learning, and proactive education. Mathematics assessments were quickly moved through various learning environments over the past year, showing that some assessment design strategies were no longer feasible if they instead centred on potential academic misconduct. Other strategies commonly used in text-based fields, however, seem to have the potential for improving both academic integrity and learning in mathematics courses. While there are some legitimate uses for digital mathematics tools and apps in STEM fields and mathematics courses, students may use them, along with 'homework help' sites, for cognitive offloading. Connection to future careers at the assessment, course, program, and institutional levels will help students contextualize the importance of academic integrity. From the perspectives of students studying mathematics, wellness may be affected by the reactive or punitive use of academic misconduct identification methods such as e-proctoring. Instructional practices focused on increasing student self-efficacy and reducing mathematics anxiety may also help students to learn with integrity.

1. Introduction

The organization of academic integrity at our college began in late 2014, with collaboration between the academic support centre – the Learning Curve – and the Library. Staff in both areas often worked with the same students on improving the same academic skills, and noted that in many cases, misconduct involving plagiarism could be prevented through assignment design. For example, instructions on using APA style citing were not always consistent with related college resources.

At the same time, some staff in each area began to delve into academic integrity research and literature. It described the culture which the college was in as being one of 'rule compliance' – that is negative, punitive, and reactive (Bertram Gallant, 2008). Along with this was citation obsession that marginalized non-text disciplines, and a sense that academic misconduct was solely a student problem.

Steps were taken to move towards a culture of academic integrity – one which is positive, proactive, educative, and supportive (Bertram Gallant, 2008). This included focusing on the purpose of citing rather than the mechanics, as it was felt that this type of academic skill could be developed at the college but be better mastered at a university. This new focus was communicated to students in both classroom and individual settings, as well as professional development sessions for faculty and administration.

A shift was also made to begin exploring avenues for academic misconduct in assessments. This included offering professional development and assistance in scaffolding, engendering higher order thinking skills, and individualizing assignments, to name but a few (see Ellis et al., 2020). The partnership between the Library and Learning Curve expanded throughout the larger Learning Commons support unit, which now contained Faculty Development Coaches and the Test Centre. Online guides for both students and faculty, containing very similar information, were created and consistently updated as new research and evidence was published. By 2019, an academic integrity advisory committee was formed by the Learning Commons. Comprised of over twenty stakeholders from five college campuses and a dozen different service areas and academic programs, it intentionally included instructors from mathematics and other non-text based fields. The group initiated a revision of the college's academic integrity policy and related processes, which now include examples related to mathematics not seen in the previous policy. Professional development sessions were offered at existing college events, but also in partnership with the provincial Manitoba Academic Integrity Network, and alignment with the International Center for Academic Integrity.

Further anchoring of academic integrity in teaching and learning was guided by several frameworks, with two combined into the graphic below (also see Bertram Gallant, 2016). From the UK's Quality Assurance Agency for Higher Education (2020) came the cycle of education, prevention, detection, and administration. This represented large amounts of proactive education for all stakeholders which the Learning Commons led, as well as their support in preventing opportunities for academic misconduct to occur, primarily through assessment design strategies, proactive student advising, and blocking access to contract cheating sites on college servers. Professional development was also offered in methods for detecting academic misconduct, though the term was changed to 'identification' based on a recommendation by Eaton (2021). Finally, the administration of policy would be improved through the advisory committee's revisions to the related policy. The inverted pyramid shape of the figure was based on that of Stephens (2016), in that the more education on academic integrity and prevention/reduction of academic misconduct which happened, the less identification would be needed, and fewer students would be in a situation in which the policy would need to be administered (Figure 1).

2. Instructional considerations

While there is no 'magic bullet' that will prevent academic misconduct in all assessments, there are things that can be done in order to help reduce it at a program or school, course, or assessment level. By addressing academic integrity at the education and prevention levels, academic misconduct can be reduced and instructors can spend less time identifying potential signs and administrating cases. By educating students about academic integrity

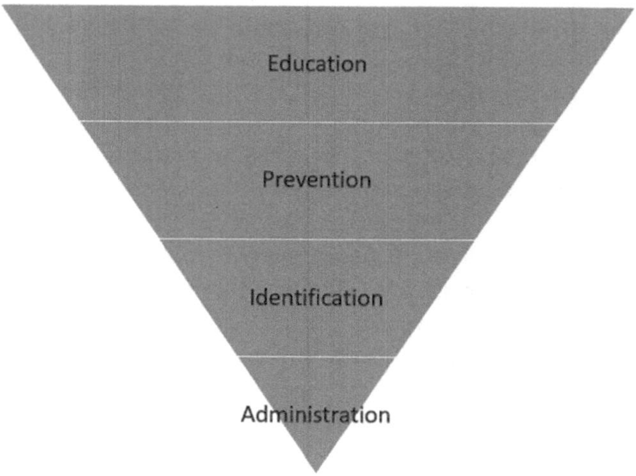

Figure 1. AI inverted pyramid.

and its risks, carefully designing assessments, and addressing academic issues when they come up, instructors can support it in the mathematics or STEM classroom.

The first tool that instructors have for supporting academic integrity is education. Even students who are comfortable with their institution's policy may not know how to contextualize it into non-prose fields (Simon, Cook et al., 2014). It is therefore important for instructors to contextualize academic integrity within their course and program. The societal safety implications of engineers and nurses with accurate mathematical skills are two examples used by the Learning Commons.

Student instruction on academic integrity can fit within different contexts and have different forms. At an institutional level, an academic integrity unit in a skills course can help to educate students about the values and importance of academic integrity during their time in school and after graduation. An example of this at the college is the academic skills course, College Foundations, which is taken by all first year students and introduces them to the values of academic integrity as well as case studies which apply to their field. In designing this content, the Learning Commons was purposeful in including examples based on mathematics and fields which study it. At the course level, a lesson could ensure that students know how these values apply to their course work and discuss the difference between collaboration and collusion in the context of their course. At the assessment level, instructors can discuss with students the expectation of the assessment, the goals and the purpose of the assessment, and what resources they are expected to use to complete the assessment. Pedagogical statements addressing specific academic integrity parameters were designed by the Learning Commons and offered to instructors for implementation. Just in time resources for students and appropriate ways to get help from the college can reduce the chances that students turn to contract cheating sites or collusion. Again, care was taken to collaborate with instructors and administrators working in fields using mathematics in order to generate consistency and inclusion in the campus culture of academic integrity.

Another opportunity for education is around the use of mathematics apps. As many students will know of mathematics apps such as PhotoMath, Wolfram Alpha, and OneNote Math, it is better to guide the narrative around the apps than try to hide them from students or, perhaps worse, ignore them. Among the biggest academic misconduct concerns with mathematics, apps are not only that they provide answers, but their abilities to solve uploaded photo problems and give step-by-step solutions. Instructors can have a conversation with their class about these tools within the context of their course and discuss how they can be used while still meeting the learning outcomes, and at what point an app is inhibiting learning. In one study, students completed exercises in order to understand how to use the work of others with integrity in a visual field (Simon Cook et al., 2014). A similar process could be used with mathematics students to promote understanding of what academic integrity and academic misconduct look like in this field. See Appendix for more information on the capabilities of some common mathematics apps.

When giving students an assignment, there are many things that instructors can do to both ensure that the assessment is a valuable learning experience for students, and help prevent and reduce academic misconduct. Students must have an understanding of the purpose of the assessment – including the skills and knowledge that they are applying. It is also important that students understand the relevance of the skills and knowledge to them as practitioners in their field. This can be achieved by making authentic assessments that mimic real-world tasks. Authentic assessments have the added advantage of being harder to cheat using mathematics apps.

For tests and exams, there are many measures that instructors can implement to prevent misconduct, often discussed in text-based disciplines but also applicable to those using mathematics. Avoiding commercial test banks, or including calculated portions (if testing online) and altered texts are good basics. Multiple versions of assessments, randomized question order, and questions written in ways that produce natural variation in student answers were all recommendations made by the Learning Commons specifically for mathematics instructors.

One measure which was initially used in the move to emergency remote learning was not allowing students to go back once an answer was given in an online quiz, test, or exam. While this was widely recommended as a way to reduce and prevent certain types of academic misconduct, it also contravened common test-taking strategies, and was anecdotally shown at the college to increase anxiety among many students. By the end of the 2020/2021 academic year, this measure was no longer recommended by the college's Learning Commons.

3. Student perspective

From the perspectives of students studying mathematics, wellness may be affected by the reactive or punitive use of academic misconduct identification methods such as e-proctoring (Eaton & Turner, 2020). Advising approaches and instructional practices, focused on increasing student self-efficacy and reducing mathematics anxiety, may help students to learn with integrity. The college uses a proactive advising model. The goal of this model is to proactively identify and support areas of student need. This is achieved through intentionally designed interventions that connect with students and educate them on supports available, before issues arise. This proactive advising model is instrumental in supporting students to learn with integrity. Proactive interventions equip students with a toolkit for addressing challenges and issues, sometimes even before students realize there is an issue. This approach may be particularly helpful in supporting what the Tertiary Education Quality Standards Association (2020) refers to as 'cheat-curious' students. 'Cheat-curious' students may consider cheating, depending on a myriad of factors. Through proactive advising students can be nudged towards learning with integrity and equipped with the knowledge, skills, and attitudes to avoid academic misconduct. Nudges help guide students and encourage them to access resources, while maintaining student agency. This advising model aligns with the prevention stage of the inverted pyramid figure shown earlier.

One example of the college's proactive approach to advising is the Math Check program. Math Check is an early identification and intervention program for college students learning mathematics. The program is a partnership between academic program staff and the Learning Curve, the college's student academic support centre. The program, started in 2013, is designed to identify early on the signs of difficulty with mathematics, and offer tailored supports. The program aims to increase the probability of student success in mathematics courses by identifying issues early and educating students on the supports available to them. To develop this program, the Learning Curve collaborated with program instructors to develop ten-question tests that assessed essential mathematics concepts in the program. Next, supports were mapped out so that students would know where and how to get help based on the results of the assessment. For example, students requiring additional support with fractions were directed to mathematics support drop-ins or appointments with the Learning Curve. Lastly, a presentation was developed that outlined mathematics requirements in the program, academic support services for students, and strategies for success.

A Math Check presentation is delivered in-class, by the Learning Curve, within the first week of classes. Presentations are either pre-scheduled by the school as part of new program intakes, or by instructors themselves. During the presentation, students write and then self-grade the assessment. Students are guided through interpreting the results and identifying supports that align with areas of need. The tone of the presentation is, like the academic integrity culture at the college, proactive, educative, and supportive. Students are encouraged to be proactive learners who take responsibility for their learning by accessing supports. Seeking help for mathematics is normalized and encouraged. Instructors can later contextualize and contrast pro-active help-seeking with the misuse of mathematics apps, contract cheating and other misconduct. The assessments are collected at the end of class, with no identifying information such as names, for the instructor and Learning Curve staff to review. This provides insight into class readiness and can inform group-based interventions such as group tutoring and advising. The Learning Curve also compiles data on student requests and access of mathematics support from professional staff and peer tutors. This provides a snapshot of how and when students seek support for mathematics, and for what concepts. This data, alongside with Math Check assessments, can be used to proactively implement scalable mathematics supports such as study groups and group tutoring in response to areas of need. This data has also been useful in developing just-in-time resources, such as handouts, study guides, and practice worksheets.

The proactive interventions employed at the college are designed to increase student self-efficacy, 'an individual's "I can" or "I can't" belief' (Siegle & McCoach, 2007, p. 278). All students can learn mathematics, and college staff are explicit in telling this to students. Anxiety about mathematics is normalized and evidence-based strategies for addressing mathematics anxiety, 'a feeling of tension, apprehension, or fear that interferes with math performance' (Ashcraft, 2002), are employed. It is understood that some students may avoid mathematics, especially if they are anxious about it, believe they cannot learn mathematics, or have previous negative experiences with mathematics. Avoiding learning mathematics due to anxiety or the use of mathematics apps or 'homework help' sites for 'cognitive offloading' (Dawson, 2021), undermines learning and reinforces the misconception that some people can't learn mathematics. By using proactive interventions to disrupt cognitive offloading, challenges in learning are reframed as part of the process of learning that can be overcome with specific strategies.

Nudging students towards legitimate and approved supports, such as the college academic support centre or online academic integrity guides, may be especially important given the noticeable increase in student use of 'outside help' during the COVID19 pandemic (Lancaster & Cotarlan, 2021). Students may not be aware of the inherent risk of breaching college policy by using contract cheating sites and mathematics apps during assessments. Students may also not be aware of college services to support their success, perhaps even more so in the emergency shift to remote instruction. Proactive advising interventions, such as Math Check, educate students about learning with integrity in a contextualized and intentional way. Providing proactive support to instructors, who are also dealing with the stress of a new learning environment, indirectly benefits students and aligns teaching and learning at the core of academic integrity culture.

4. Takeaway

Many strategies which help promote academic integrity and prevent or reduce academic misconduct such as plagiarism and contract cheating in text-based disciplines can also be applied to mathematics contexts by collaboration between institutional support units and mathematics instructors.

5. Conclusion

Using a centralized and coordinated approach towards academic integrity, colleges and universities may find a multifaceted anchor for their stakeholders to not only survive this sudden change in learning environments, but to thrive. While much of the literature focuses on text-based disciplines, it may be applied to mathematics and other STEM fields. In particular, a focus on teaching and learning rather than academic misconduct, and being positive, educative, supportive, and proactive with academic integrity efforts helps to improve the learning environment for all stakeholders while reducing potential academic misdconduct. Leveraging institutional supports and the concepts of student self-efficacy and mathematics anxiety within the academic integrity context also helps to make it more inclusive of mathematics and other STEM fields.

Disclosure statement

No potential conflict of interest was reported by the authors.

References

Ashcraft, M. H. (2002). Math anxiety: Personal, educational, and cognitive consequences. *Current Directions in Psychological Science*, *11*(5), 181–185. https://doi.org/10/cw57wn

Bertram Gallant, T. (2008). *Academic integrity in the 21st century: A teaching and learning imperative.* Springer.

Bertram Gallant, T. (2016). Leveraging institutional integrity for the betterment of education. In T. Bretag (Ed.), *Handbook of academic integrity* (pp. 979–993). Springer.

Dawson, P. (2021). *Defending assessment security in a digital world: Preventing e-cheating and supporting academic integrity in higher education.* Routledge.

Eaton, S. E. (2021). *Plagiarism in higher education: Tackling tough topics in academic integrity.* Libraries Unlimited.

Eaton, S. E., & Turner, K. L. (2020). *Exploring academic integrity and mental health during COVID-19: Rapid review.* https://doi.org/10/gjj2mm.

Ellis, C., van Haeringen, K., Harper, R., Bretag, T., Zucker, I., & McBride, S. (2020). Does authentic assessment assure academic integrity? Evidence from contract cheating data. *Higher Education Research & Development*, *39*(3), 3. https://doi.org/10.1080/07294360.2019.1680956

Lancaster, T., & Cotarlan, C. (2021). Contract cheating by STEM students through a file sharing website: A Covid-19 pandemic perspective. *International Journal for Educational Integrity*, *17*(1), 1–16. https://doi.org/10.1007/s40979-021-00070-0

Quality Assurance Agency for Higher Education. (2020). *Contracting to cheat in higher education: How to address essay mills and contract cheating.* https://www.qaa.ac.uk/docs/qaa/guidance/contracting-to-cheat-in-higher-education-2nd-edition.pdf.

Siegle, D., & McCoach, D. B. (2007). Increasing student mathematics self-efficacy through teacher training. *Journal of Advanced Academics*, *18*(2), 278–312. https://doi.org/10/dpdxtx

Simon Cook, B., Sheard, J., Carbone, A., & Johnson, C. (2014). Academic integrity perceptions regarding computing assessments and essays. *Proceedings of the Tenth Annual Conference on International Computing Education Research* (pp. 107–114). https://doi.org/10.1145/2632320.2632342.

Stephens, J. M. (2016). Creating cultures of integrity: Multilevel intervention model for promoting academic honesty. In T. Bretag (Ed.), *Handbook of academic integrity* (pp. 995–1007). Springer.

Tertiary Education Quality Standards Association. (2020). *Academic integrity in Australian higher education: A national priority.* https://www.teqsa.gov.au/sites/default/files/academic-integrity-infographic.pdf?v=1574919157.

Appendix. Math app comparison

App name	Phone app	Desktop accessible	Upload photo problems	Provides answers	Step-by step solutions	Generates quiz	Graphing	Examples	Links to video lessons	Specialized calculators	Textbook solutions	Other
Photomath	Yes	No[a]	Free	Free	Free	No	Free	No	Pay	No	Pay	
Mathway	Yes	Yes	Free	Free	Pay	No	Free	Free	No	Limited Selection	Pay	Chegg Service[b]
Wolfram Alpha	Pay[c]	Yes	Pay	Free	Pay	Pay	Free	Free	No	Extensive Selection	No	
OneNote	Yes	Yes[d]	No	Free	Free	Free[e]	Free	No	No	No	No	
Microsoft Math Solver	Yes	Yes	Yes	Free	Free	Free	Free	Free	Free	Limited Selection[f]	No	

[a] Photomath is available on a Desktop computer through an Android emulator.
[b] As a Chegg service, this app promotes other Chegg services when the student uses it.
[c] Wolfram Alpha is accessible online via any device for free; Wolfram Alpha also has specialized calculator Web Apps with the Pro version.
[d] The Math function is available by default with some versions of OneNote, and as an add-in with others.
[e] OneNote uses MS Forms to generate a practice quiz.
[f] MS Math Solver has some advanced mathematics calculators, such as a calculus calculator and a matrix calculator, but does not have calculators for other fields such as chemistry, finance, etc.

Individualized summative assessments as used during COVID-19

Nathan Clisby and Antony Edwards

ABSTRACT

In this classroom note, we outline a system of assessment used by the authors since 2020 to deliver individualized summative assessments to students from first- and second-year mathematics courses. Our system comprises three modular components allowing a mix-and-match of different technological approaches and mathematical question types. First is a question generation module that generates appropriate variables and question syntax, second is a delivery module to send out the individualized assessment to students, and third is a marking module to generate worked solutions and final answers for markers. The key benefits of these assessments are an ability to incorporate individualized authentic elements into assessments, to allow access to technology that would be impractical for an invigilated exam setting, whilst overall reducing the likelihood – but increasing the ease of detection – of academic misconduct and contract cheating, compared with other non-invigilated assessment protocols.

1. Introduction

In this classroom note, we present our platform-agnostic approach to generating individualized student assessments. This system allows complete flexibility for design, delivery, and feedback unconstrained by a single technology choice. It has been used at our university across multiple mathematics units, year levels and assessment types: from project-based assessments already using this approach prior to 2020 (Edwards, 2018), to high-stakes mid-term tests and final exams included due to COVID-19 restrictions.

In common with the tertiary sector as a whole, over the last ten years at Swinburne University of Technology there has been a move towards a blended online approach to teaching and learning in tertiary mathematics classes, with the introduction of auxiliary video materials in the form of MathsCasts (Loch et al., 2011), automated lecture recordings, and live-streaming of classes accompanied with in-class polling (Edwards, 2019). In contrast to online teaching delivery, until the COVID-19 pandemic necessitated all assessments to be online, summative assessment in mathematics usually comprised of one or two invigilated mid-term class tests and a final invigilated exam. There is evidence that such a reliance on assessment via high-stakes invigilated tests disadvantages women and students from marginalized (or non-traditional) backgrounds (Jameson & Fusco, 2014; Oswald & Harvey, 2000), and does not assess what employers are looking for when hiring graduate employees (Yorke, 2006).

Exceptions to these forms of summative assessment were made by the authors prior to 2020 to include those in the form of take-home projects where either a time limit was inappropriate or where external tools and information were required for the assessment to be meaningful or authentic to the question studied (Edwards, 2018). Such assessments were written for domain-specific service mathematics units such as those for students studying a Biology or Aviation Major and would be considered *authentic* in the sense of Ellis et al. (2020). We do note that although there is no consensus as to what exactly constitutes an authentic assessment (Frey et al., 2012), when appropriately embedded such assessments appear to be instrumental in helping students become good professionals (Villarroel et al., 2020). It is within such projects that our present process of developing individualized assessment briefs originated and which, in 2020, due to COVID-19 restrictions, was rolled out to encompass all tertiary mathematics assessment in the authors' units, including replacing high-stakes invigilated tests and exams. This classroom note concentrates on our system for authoring, delivering, and marking assessment, rather than the content of individual assessments. Finer points and various tips and tricks are illustrated by presenting appropriate examples and in the Discussion section.

2. Aims of the assessment delivery system

Individualized assessment in mathematics is constrained by a lack of flexibility in three aspects: assessment content, means of delivery to students, and the mechanism by which students submit answers. Learning and Teaching Interfaces (LTIs) set up by textbook manufacturers, paid subscription services, or open-source communities will allow different levels of control on these aspects when used in conjunction with a university's Learning Management System (LMS). At one extreme, a system will consist of a selection of pre-authored questions that can be delivered to students to answer on a proprietary software interface at a time of the instructor's choosing. Such an offering usually allows instructors no flexibility to modify questions, to change how students submit answers, or to modify how these answers are stored. Systems with sufficient flexibility to allow users to author their own questions must provide an interface for the authoring which may involve an instructor learning a suitable programming language e.g. Maxima in the case of STACK (Sangwin, 2015) or JME for Numbas (Foster et al., 2012). Both pre-authored and user-authored systems have the benefit that an LTI provides an all-in-one system of question generation, student access, question marking, and feedback.

However, there are limitations in using an all-in-one approach. As noted by the creator of a popular computer-aided assessment tool in mathematics – STACK (Sangwin, 2015) – questions that are authored, implemented and marked by a computer algebra system may have 'fundamental problems of ... (i) question distortion, (ii) assessment only of lower-order skills, and (iii) strategic learning [of content].' (Sangwin, 2013, p. 987–988). We note there has been much progress in this area, in particular in the areas of assessing students' understanding of proof (Bickerton & Sangwin, 2021), but it is still fundamentally the case that for higher-stake summative assessment, where approaches to answering questions are often as important as final answers, such systems are inflexible and unsuitable.

Wishing to allow complete flexibility for question authoring, and with the starting position that students will eventually be submitting written answers that would be manually marked, we developed a system that would not be reliant on a specific LTI or programming language. We wished to use whichever tool felt the most appropriate at any given time, whilst ensuring that others could follow the procedure we used (e.g. in the case of handing across to a colleague to teach). As assessments would be delivered for students to complete in their own time, we wanted to produce individualized assignments in which each student had their own (or at least a randomized set of) variables in a question to reduce the likelihood that students would see value in copying answers (c.f. Manoharan, 2016). Another requirement of the system was that, aside from the time for initial setup, we needed our assignments to be scalable for deployment and marking in large classes, with at most a small constant increase of marking time as the cohort size grew.

3. Assessment-independent modular structure

The key feature of our system is to completely decouple the generation of questions, the delivery to students, and marking, and feedback (see Figure 1). By decoupling, we allow flexibility for an instructor to deliver the assessment that they deem appropriate for their students (within the proviso that it be an online assessment) using tools that they are familiar with. Within this overarching approach, in different assessments, we have used Microsoft Office, MATLAB, Bash scripts in a Linux environment, and the programming language Julia at the various stages. Such an approach was essential during COVID-19 teaching semesters in 2020 when there was insufficient time to train in a new language or software approach. The best tool for a job is the one that you can use most effectively.

3.1. Generation of questions

For structured activities, individualized questions require parameters to be embedded. Before this can happen, a question frame must be authored allowing for variable-based input, represented in the examples below by $\boxed{\text{boxed text}}$. Due to an increased potential to identify academic misconduct later, it is worth including esoteric phrases at this stage. Such questions may involve a standard setup with a quirky feature added, this itself may be a variable that changes alongside the in-question mathematics variables:

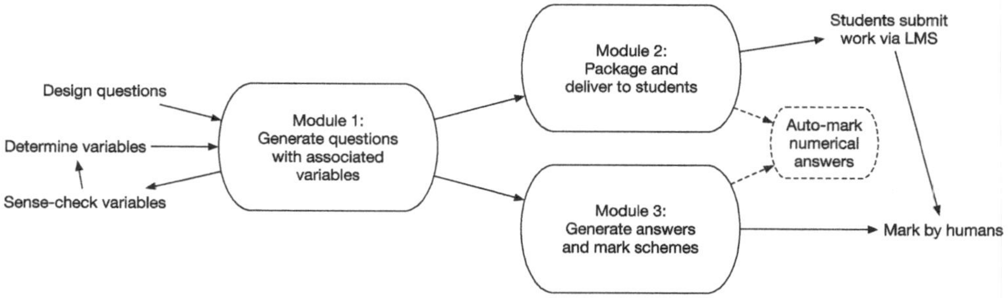

Figure 1. A flowchart describing the relationship between each stage of the assessment-independent structure. Dashed lines indicate possible extensions of the structure not yet realized by the authors.

A $\boxed{\text{LEGO monorail}}$ has a straight section of track along the negative x-axis. At the origin, the track connects to a transition curve with the equation $y = \boxed{3}\, x^3$ which connects to a circular track at (x, y) where $x = \boxed{1}$ cm. Find the radius of the circular track.

As students were uploading typed or handwritten answers, more flexibility in question types was possible. Following the pedagogical approach of Watson and Mason (2006), we found it particularly effective to ask students to use their own examples of objects subject to randomized constraints:

Give an example of a vector $\vec{w} = (a, b)$ where $|\vec{w}| = \boxed{a}$, or explain why it is impossible.

We also developed more general free-form questions where students gathered their own objects with which to construct an analysis, subject to the variables in the question:

Collect data for a recently completed flight between $\boxed{\text{Melbourne}}$ and $\boxed{\text{Perth}}$ on flightaware.com and model the direction at the flight's midpoint by a 2D unit vector. Why is this unit vector unlikely to be in exactly the same direction as the midpoint of the great circle connecting these cities?

A similar approach was taken in a statistics component where we asked students to perform analyses and formulate appropriate hypotheses regarding temperature and rainfall data at locations around Australia. Each student was emailed a unique pair of locations and provided with access to a CSV file with an extensive climate dataset collated from the Bureau of Meteorology, thus ensuring that their results would be unique to them.

To produce suitable rules for variables, it can be helpful by starting with the format of the answer, a process described well by Sangwin (2013, ch 4). Once the question frame is ready, the variables should be coded. Various approaches here are sensible, including manually listing items and using a random selection (Excel: `index` and `randbetween` commands, MATLAB: `datasample`), or using random selections based on logical rules (Excel and MATLAB: `rand`). For our Julia questions we looped over all possible combinations of parameter values, then subsequently randomized the assignment of these questions to students. Depending on the complexity of the question more advanced logic may be necessary, but this can be simplified by 'pasting values' in Excel to remove issues where one random choice precludes the selection of another (e.g. two different zeros of a function), or using set differences in MATLAB. Finally, a CSV file is produced (Excel: `Save As`, MATLAB: `array2table` then `writetable`).

3.2. Delivery to students

In this stage variables are fitted into questions and then sent to students. This can be achieved via a mail merge in Microsoft Office, insertion of variables into LaTeX code, or direct insertion into an LMS where allowed.

For a mail merge approach, full questions can be authored in Word with variables added as mail merge fields (including in equations). Careful checking is required to confirm that variables are correctly formatted after the merge; field codes (Alt-F9) should be used to correct formatting (useful commands are # for decimal place rounding or currency options and @ for various date settings). If an Adobe Acrobat plugin is installed, individual PDF files can then be generated and automatically sent to each student email address during the mail merge.

If a university allows SMTP access to its email server, other approaches are possible. For instance, using the programming language Julia variables can be entered directly into individualized LaTeX documents, which can then be compiled into pdf files and emailed to students via appropriate scripts (we used Bash scripts in a Linux environment). One benefit of this approach is that it allows one to leverage the high-quality mathematical typesetting of LaTeX documents.

3.3. Marking and feedback

We created marking guides for each human marker based on the randomized variables. Our goal was to make the marking process as efficient, fair, and consistent as possible. Note that the nature of the questions and the free-form responses of students necessitated a human marker. It would be possible to automate some aspects of the marking process within our system (see Figure 1), in particular numerical correct–incorrect answer types, but for us this would have involved students submitting answers to a system outside our LMS. Furthermore, given that human markers were necessary to assess free-form question parts, the relative ease of marking numerical answers at the same time would mean that the efficiency gain from a partially automated marking system would be minimal.

To create the marking guides we took the variable list and constructed milestone calculation points. The number of milestones depended on the complexity of the question; first-year calculus questions can often be checked by markers without an answer for comparison, while more sophisticated second-year questions may require half a dozen milestones to ensure that the marker can check the students' work without requiring laborious calculations. With additional time allocated to the development of the marking system, it would be possible to have a separate milestone entry system for students where marks could be given for milestones, then verified by human markers. The human marker could then concentrate on assessing the mathematical communication skills of students.

Student submission of answers via an LMS also allowed rudimentary checks and measures such as automated Turnitin reports to be generated to identify any highly similar (typed) solutions. We found a high rate of false-positives in this approach; Turnitin tends to weigh more heavily question sentences over displayed mathematics yielding high similarity scores for any student who included the question text before their answers, for instance.

4. Discussion

As with any assessment containing variables that may take different forms for different students, attention must be paid to ensure variables give every student a meaningful question of an appropriate standard. A good way of doing this is first to test edge cases, then to sense check 10–15 randomly generated questions, taking into account that it is normal for fringe cases to throw up unexpected pairings where several variables are dependent on each other.

Designing effective parametrized questions can be challenging, especially if one wishes for the questions to be unique to each student. For a structured assessment for a second-year maths course, we asked students to calculate Fourier series for an individualized piecewise linear function. Choosing a family of functions and associated parameters which were (a) unique, (b) of the same difficulty, and (c) pedagogically useful, was quite challenging. On reflection, the challenges of accommodating (a) and (b) led to the questions being more fiddly than was ideal for pedagogical utility. In particular, it may sometimes be necessary to sacrifice uniqueness in parameter choice, and instead have a variety of questions. One consequence of parametrized questions is that simple choices of parameters, which might be desirable for confidence-building questions, must generally be excluded. For instance, it is unfair to ask some students to compute $\int \sin x \, dx$ and other students to compute $\int \sin 7x \, dx$, as the latter integral is more challenging. There are other reasons for excluding the former: if a mail merge is used, it would result in the unwieldy $\int \sin 1x \, dx$.

Our markers reported that the additional time to mark a randomized question was negligible. This was partly because care was taken when generating marking guides to ensure that the order of solutions matched the order in which the LMS presented assessments. Furthermore, although the variables changed between students the context of the answers did not, so the usual speeding up of marking occurred as the ways in which students could err within a question became more established. Some markers opted themselves to construct more sophisticated guides that were then used by other markers. We note that it would be desirable to quantify the extra time required for marking individualized assessments, but as we did not implement any control group the best we can do is rely on qualitative feedback from markers.

The combination of esoteric phrases and randomly generated mathematical variables allowed for the identification of individual instances of academic misconduct. In one early major assessment, over 15% of students had either requested or viewed a solution from one single homework subscription website (and these were just those who did not obfuscate their online identity). As there is reasonable evidence that using individualized assessments decreases academic misconduct (Manoharan, 2016), we possibly encountered so many because it was much easier to detect via Turnitin and searches on popular homework-help websites. One approach that proved extremely useful, after we discovered that academic misconduct was occurring, was to create Google Alerts with identifiable snippets of questions. In this way, we would receive email alerts of instances of questions being posted to contract-cheating websites, typically within hours of these questions appearing. This measure significantly decreased the cognitive load, and worry, of continual monitoring for academic misconduct.

Our surprise is perhaps naive in the context of studies that found 6% (Bretag et al., 2019) and 70% (Broeckelman-Post, 2008) of students admit to some form of cheating at university, and that authentic features in assessments may actually increase instances of academic misconduct (Ellis et al., 2020). As noted by Seaton (2019), there appears to be little research in academic misconduct in tertiary mathematics, with the exception of some research on automated detection (Meuschke et al., 2017, 2019; O'Malley & Roberts, 2012). Although we observed a high rate of contract cheating in early assessments, it dramatically declined when it became clear to students that cheating could and would be detected, and once we established a dialogue around issues of academic integrity.

5. Takeaway

A system to deliver individualized assessments has been developed by the authors during the period of COVID-19 restrictions. This system incorporated flexible components of generation, delivery, and marking, so as to make it highly scalable. We have found that the risks of academic misconduct, and in particular contract cheating, can be substantially mitigated by having a dialogue with students about academic integrity, and through invigilation by monitoring the appearance of test questions online via Google Alerts. These assessments are more authentic than the time-limited in-class tests they have replaced and will continue to be used by the authors when the COVID-19 crisis ends.

Acknowledgments

We thank Emily Cook for the use of her LEGO monorail example question. The authors are happy to provide code and parametrized questions on request.

Disclosure statement

No potential conflict of interest was reported by the authors.

ORCID

Nathan Clisby http://orcid.org/0000-0001-7068-8347
Antony Edwards http://orcid.org/0000-0002-8816-5488

References

Bickerton R., & Sangwin C. (2021). Practical online assessment of mathematical proof. *International Journal of Mathematical Education in Science and Technology*, 1–24. https://doi.org/10.1080/0020739X.2021.1896813

Bretag T., Harper R., Burton M., Ellis C., Newton P., Rozenberg P., Saddiqui S., & van Haeringen K. (2019). Contract cheating: A survey of Australian university students. *Studies in Higher Education*, 44(11), 1837–1856. https://doi.org/10.1080/03075079.2018.1462788

Broeckelman-Post M. A. (2008). Faculty and student classroom influences on academic dishonesty. *IEEE Transactions on Education*, 51(2), 206–211. https://doi.org/10.1109/TE.2007.910428

Bureau of Meteorology. *Climate data online*. http://www.bom.gov.au/climate/data/.

Edwards A. (2018). *Individualised summative assignments that can be scaled sustainably*, First Year in Mathematics Group, University of Melbourne. https://fyimaths.files.wordpress.com/2018/08/edwards-2018-fyi-maths-handout3.pdf.

Edwards A. (2019). Student polling: It's the taking part that counts. In *Proceedings of the 12th Delta Conference on the Teaching and Learning of Undergraduate Mathematics and Statistics*. https://researchdirect.westernsydney.edu.au/islandora/object/uws%3A52645.

Ellis C., van Haeringen K., Harper R., Bretag T., Zucker I., McBride S., Rozenberg P., Newton P., & Saddiqui S. (2020). Does authentic assessment assure academic integrity? Evidence from contract cheating data. *Higher Education Research & Development*, 39(3), 454–469. https://doi.org/10.1080/07294360.2019.1680956

Foster B., Perfect C., & Youd A. (2012). A completely client-side approach to e-assessment and e-learning of mathematics and statistics. *International Journal of E-Assessment*, 2. https://ijea.org.uk/index.php/journal/article/view/35

Frey B. B., Schmitt V. L., & Allen J. P. (2012). Defining authentic classroom assessment. *Practical Assessment, Research, and Evaluation*, 17, 2. https://doi.org/10.7275/sxbs-0829

Google Alerts. https://www.google.com/alerts.

Jameson M. M., & Fusco B. R. (2014). Math anxiety, math self-concept, and math self-efficacy in adult learners compared to traditional undergraduate students. *Adult Education Quarterly*, 64(4), 306–322. https://doi.org/10.1177/0741713614541461

Loch B., Gill O., & Croft T. (2011). Complementing mathematics support with online MathsCasts. *ANZIAM Journal*, 53, C561–C575. https://doi.org/10.21914/anziamj.v53i0.4984

Manoharan S. (2016). Personalized assessment as a means to mitigate plagiarism. *IEEE Transactions on Education*, 60(2), 112–119. https://doi.org/10.1109/TE.2016.2604210

Meuschke N., Schubotz M., Hamborg F., Skopal T., & Gipp B. (2017). Analyzing mathematical content to detect academic plagiarism. In *Proceedings of the 2017 ACM on Conference on Information and Knowledge Management* (pp. 2211–2214). IEEE. https://doi.org/10.1109/JCDL.2019.00026

Meuschke N., Stange V., Schubotz M., Kramer M., & Gipp B. (2019). Improving academic plagiarism detection for STEM documents by analyzing mathematical content and citations. In *2019 ACM/IEEE Joint Conference on Digital Libraries (JCDL)* (pp. 120–129). IEEE.

O'Malley M., & Roberts T. S. (2012). Plagiarism on the rise? Combating contract cheating in science courses. *International Journal of Innovation in Science and Mathematics Education*, 20, 16–24. https://openjournals.library.usyd.edu.au/index.php/CAL/article/view/5803

Oswald D. L., & Harvey R. D. (2000). Hostile environments, stereotype threat, and math performance among undergraduate women. *Current Psychology*, 19(4), 338–356. https://doi.org/10.1007/s12144-000-1025-5

Sangwin C. (2013). *Computer aided assessment of mathematics*. OUP.

Sangwin C. (2015). Computer aided assessment of mathematics using STACK. In *Selected Regular Lectures From the 12th International Congress on Mathematical Education* (pp. 695–713). Springer.

Seaton K. A. (2019). Laying groundwork for an understanding of academic integrity in mathematics tasks. *International Journal of Mathematical Education in Science and Technology*, 50(7), 1063–1072. https://doi.org/10.1080/0020739X.2019.1640399

Villarroel V., Boud D., Bloxham S., Bruna D., & Bruna C. (2020). Using principles of authentic assessment to redesign written examinations and tests. *Innovations in Education and Teaching International*, 57, 38–49. https://doi.org/10.1080/14703297.2018.1564882

Watson A., & Mason J. (2006). *Mathematics as a constructive activity: Learners generating examples*. Routledge.

Yorke M. (2006). *Employability in higher education: What it is – what it is not* (Vol. 1). Higher Education Academy.

Learning at a distance: can at-home activities measure up?

Jessica Gehrtz, Raquel Vallines Mira, Christopher Duffer and Priya V. Prasad

ABSTRACT
Due to the global pandemic, many traditionally face-to-face courses had to transfer to online teaching. We, like many others, were faced with the challenge of replicating online our purposefully-designed mathematics content courses for elementary teacher candidates (TCs). In doing so, our aim was to preserve our pedagogical commitments of collaborative learning and equitable participation. In this paper, we describe a measurement activity that we redesigned in the rapid transition to online teaching. This activity exemplifies how the constraints of remote teaching opened new opportunities for maintaining our pedagogical commitments in different and richer ways. We also highlight two takeaways from this work: (1) creating space for students to explore introductory activities at home can lead to richer class discussions and more equitable participation, and (2) allowing for students to share their work in an online environment can lead to more creative approaches and presentations, as well as opportunities for students to share in-progress thinking before whole class discussions.

1. Introduction

In the transition to online instruction that occurred in response to the COVID-19 pandemic, many instructors were faced with the challenge of adapting courses for an online format. Our pedagogical commitments guided much of the lesson development process for our in-person elementary content courses. As we shifted to online synchronous instruction, it was important for us to preserve the footprint of our commitments in the newly designed lessons. Specifically, it was important for us to create a classroom environment that (1) engaged teacher candidates (TCs) in collaborative learning and (2) promoted equitable participation. In this paper, we describe our pedagogical commitments and focus on one lesson that we modified for online instruction. Although this modification was made in order to fit the lesson into the constraints of online teaching, the result exceeded our expectations and impacted how we will implement this activity in the future.

Our first central pedagogical commitment for our elementary content courses is involving TCs in collaborative learning. Every lesson we have designed for this course requires TCs to work together, discuss concepts, and find solutions to group-worthy tasks. Lotan (2003) describes group-worthy tasks as open-ended tasks that contain multiple entry points so that every student has something to contribute, and require interdependence and collaboration among small groups of students. Research has shown that working on group-worthy tasks is a key way for students to engage deeply and conceptually with mathematics (Boaler, 2015; Cohen & Lotan, 2014; Lotan, 2003).

Our second central pedagogical commitment of promoting equitable participation in both small- and whole-group discussions goes hand-in-hand with collaborative learning. Since participation in class discourse is an important resource for learning, equity in collaborative learning can be defined as 'a fair distribution of opportunities to learn or opportunities to participate' (Esmonde, 2009, p. 1010). In order to promote more equitable participation, we work to create a classroom environment where TCs feel safe to make mistakes and where incomplete and in-progress ideas are valued as part of class discussion and the learning process. Since TCs do not need to have fully formed and perfect answers before sharing, the risk of sharing is reduced and participation in discussion tends to increase (Jansen, 2020).

2. In-person version of the measurement activity

Here we describe one lesson for our elementary mathematics content course that exemplifies our pedagogical commitments. This lesson, which we will refer to as the Measurement Activity, serves as an introduction to a module on measurement. The primary objectives of this lesson are to develop TCs' understanding of (1) the process of unit iteration when measuring and (2) the inverse relationship between size of unit and number of units needed to measure something (e.g. length, area). This module is one of several modules and lessons that we designed, assessed, revised, and reassessed until we felt it addressed our stated student learning goal.

For the in-person implementation of the Measurement Activity (Figure 1), TCs worked in small groups and were each given a strip of construction paper (3 or 4 strips total depending on group size). Since our elementary content courses generally have a maximum enrolment of 30 TCs, there are typically 8–10 small groups per section of the class. For some groups, the strips were 9 inches long and for others 1.5 feet long (all strips were 3 inches wide). TCs were asked to use these strips to measure, with as much precision as possible, the length of the same architectural feature of campus that measured at least 10–15 feet. Since TCs were told they would only have two minutes to do the actual measuring, they had to strategize ahead of time about the most efficient way to do this. We encouraged TCs to think about iterating the given lengths or concatenating them to form longer units of measurement. TCs were also given access to tape and scissors, but not given any directions on how or whether to use these resources. After finishing their measurements, groups recorded their results on the board and the instructor facilitated a whole class discussion about groups' strategies. Often, this discussion included opportunities for groups to compare strategies, notice the variation in their results, and hypothesize ways to achieve greater measurement precision.

The instructors who designed and implemented this in-person activity consider it to be an excellent illustration of our pedagogical commitments. The Measurement Activity requires collaboration as groups strategize the best way to combine their strips of paper of various lengths and use their tools to measure the architectural feature. Further, this activity sets the stage for equitable participation. Since this is the introductory activity for

Find the distance between point A and point B.

Each student in your group will be given a strip of paper that is 9 inches long, tape and scissors. You will be asked to use the strips to measure, as accurately as possible, the distance between points A and B, as shown in the picture. You will have only 2 minutes to complete the measurement.

Before you start measuring, your group will have some time to develop the most efficient strategy. You will also get some time after the measurement to finalize your answer. Keep in mind that the goal is to be as accurate as possible.

1. Record your groups' measurements in the table shown below. You do not have to use all the columns on the table. Use only the ones useful to your group.
2. If you have not already done so, fill out at least 3 columns of the table shown.

Table of Measurements

Number of 9-inch strips	Number of 1.5-feet strips	Number of inches	Number of feet	Number of yards	Number of ________

Figure 1. Teacher candidates' worksheet for in-person version of the measurement activity.

a module on measurement, it does not require familiarity with standard units of measurement and it allows for diverse strategies. In fact, non-standard strategies for measuring actually tend to create opportunities for richer discussion and to surface implicit assumptions about measurement. However, our strategies for promoting equitable participation remain a work-in-progress; we have found that for in-person implementations of this activity that it is possible for TCs to be part of the small group and yet not be actively involved in the idea generation or measurement.

3. Online version of the measurement activity

In the rapid transition to online instruction, we were faced with modifying our carefully-designed Measurement Activity for use in the fully online, synchronous version of this course. For the online version, TCs still had to measure an architectural feature. Instead of requiring collaboration with their peers on campus, we asked individual TCs to measure an architectural feature in their home and encouraged them to involve family members or friends. The TCs were instructed to cut a strip of paper (or some other material) that was no longer than the distance from their elbow to the tip of their fingers, and no shorter than the distance between their elbow and wrist to use for measuring the space. The purpose of this was to preserve the non-standard unit of measure that we provided in the in-person version of the activity.

We introduced the Online Measurement Activity, going over the instructions in Figure 2, during one of our synchronous online classes. For the Online Measurement Activity, TCs were also asked to answer the reflection questions in Figure 3, and to include pictures of their tools and implementation on a shared set of Google Slides. This was

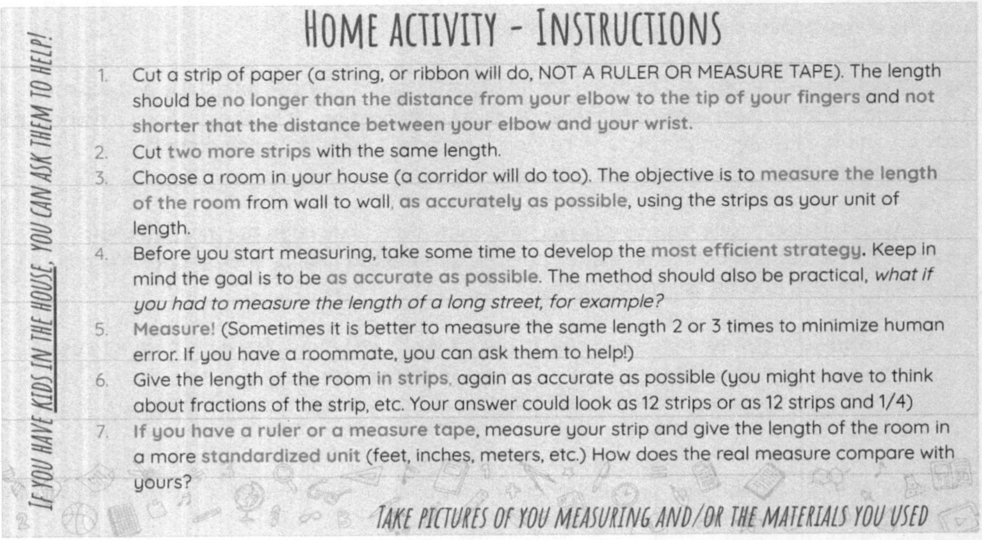

Figure 2. Instructions for online version of the measurement activity.

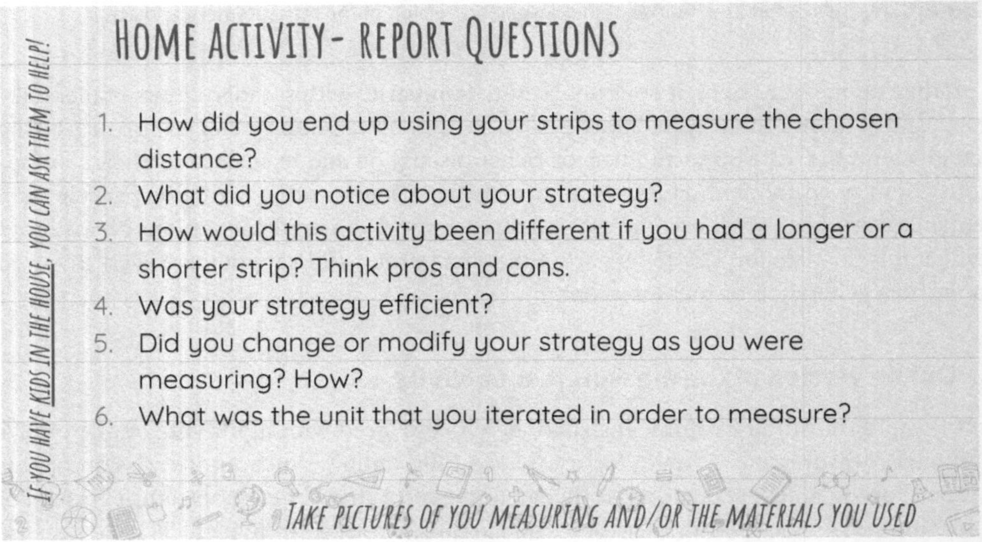

Figure 3. Questions to guide teacher candidates' reflections on the online measurement activity.

one key difference from the in-person implementation because in that version TCs were prompted with questions only after measuring the architectural feature. The Google Slides were shared with the class and all TCs had editing permissions. Hence, as soon as TCs answered a question on the Google Slides, the rest of the class had access to the in-progress work. The measurement task, along with the reflection questions, were all completed by the students asynchronously before the next class meeting.

In the next class meeting, the instructor initiated a whole class discussion about the Online Measurement Activity, asking TCs to share their process for measuring the architectural feature, their reflections about the activity, and to comment on other's strategies. TCs contributed to the discussion by talking in the video call and through the chat feature of the video call platform. During the discussion, the instructor shared their screen so that all TCs could see the Google Slides with the responses of the TC who was sharing at the time. Occasionally, the instructor would ask a specific student to share if their response on the Google Slide highlighted an idea that had not yet been discussed. Although there were a few TCs in each section that did not complete the activity, the class discussion was not limited. Many of these students still contributed by commenting on the strategies of others or sharing how they might have approached the task differently. We suspect that most students completed the activity before class because it was a hands-on activity introducing a new module and so it required little prerequisite knowledge. Additionally, throughout the course we have worked to create a classroom environment where students' thinking is valued and it is a safe space to share ideas, even if they are not fully formed or correct; this made completing the Online Measurement Activity and sharing measuring strategies low risk.

The Online Measurement Activity exemplifies how we were able to maintain (and even amplify) our pedagogical commitments in the transition to online learning. Despite the fact that TCs were not collaborating with their peers in the class for the activity, the whole class discussion following this individual work allowed for collaborative learning. The class discussion built on individual TCs' ideas and allowed them to reflect on their own experiences as well as their peers' strategies. By encouraging TCs to include their families as they engaged with the task, this also created an opportunity for collaborative learning to occur outside of the classroom.

The Online Measurement Activity created even more opportunities for equitable participation than did the in-person version. Since TCs were required to come to class with their answers to the questions on the Google Slides, almost every student had already thought deeply about the mathematical concepts, came to their own conclusions while answering the prompting questions, and were prepared to share their experiences. Further, this led to a greater diversity of strategies being shared and discussed than what had arisen in the in-person Measurement Activity. We hypothesize that this is partly because in the online implementation, the TCs did not have time to censor their initial ideas because others in the class had immediate access to their responses in the shared Google Slides. We also observed that more TCs, and different TCs than typically volunteered to share, took opportunities to participate in the discussion.

3.1. Outcomes of online measurement activity

While the online implementation of the Measurement Activity was designed to mirror the in-class activity as much as possible, there were undeniable differences in the online implementation. The Online Measurement Activity required the TCs to work on the activity outside of the classroom context, which removed the role of the instructor and the time constraints of an in-person class activity. For warm-up activities like this, where instructors plan to scaffold student ideas into a meaningful discussion, we can often underestimate the abilities of our students to do this successfully on their own and overestimate our own

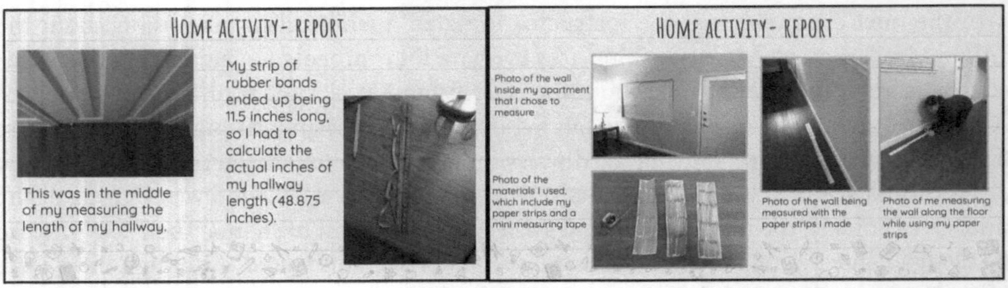

(a) (b)

Figure 4. Two teacher candidates' work for the online measurement activity.

importance in this process. We were excited to see that in the online version students came up with many important discussion points on their own. These points went beyond basic observations about the activity itself and extended into meaningful discussions about measurement in general. For example, a student at home chose to use a chain of rubber bands as their unit of measurement (Figure 4(a)). This prompted a discussion about the characteristics of units when one student pointed out that a rubber band can take on different lengths depending on whether it is stretched out or not. While we typically would not have provided this material in class, allowing the students more freedom in the online implementation led to discussions that the resources provided during in-person implementation did not catalyze.

The online implementation also presented other unique opportunities as a result of this format. Since TCs completed this activity at home, and the class meeting was online, it meant they would be submitting and presenting their work electronically. Asking students to present their answers to discussion questions and photographs of their work in a Google Slides document resulted in more creative responses, both in terms of tools used (the rubber band) and in the presentation of strategies (see Figures 4 and 5 for examples of TCs' presentations). Further, as mentioned above, we saw more equitable participation in the online implementation. It may be that students, who are typically accustomed to interacting and sharing pictures online, felt more comfortable expressing their ideas in this digital manner, especially when they had the time and space to do so outside of the pressures of class time. Another opportunity presented by the online implementation was the option to involve family members in this activity. TCs with children at home often face additional struggles that come with balancing an education and parenthood. However, especially in a class for future elementary teachers, children at home can be a powerful resource that should be brought into the classroom and can provide valuable insights into how elementary students themselves might respond to an activity like this. Figure 5 shows screenshots from a video one TC created as she worked on the Online Measurement Activity with her children.

4. Key takeaways

This task exemplifies how the constraints of online teaching opened new opportunities for maintaining our pedagogical commitments in different and richer ways. Further, the

Figure 5. Teacher candidates involved their families in the online measurement activity.

transition to online instruction has created opportunities for us to reflect on what aspects of this Online Measurement Activity we will take with us into future implementations, and to reflect on key takeaways that could be applied to other activities or contexts.

One key takeaway from the Online Measurement Activity is that creating space for students to engage individually with and explore learning opportunities at home can lead to richer class discussions and more equitable participation. Having students complete work at home is not a new concept, whether it be completing problems for homework or viewing lectures in a flipped-classroom style. We suspect that the exploratory nature of the Online

Measurement Activity is one aspect that distinguishes it from these other activities. Unlike an activity that is typically designed to be completed at home, this activity was designed to introduce a new topic and get students 'hooked', and is commonly employed in the classroom. However, since we saw such success with the online implementation, we will continue to have TCs complete this activity at home when returning to the classroom as it provides space for them to engage with the content individually, before class discussion, without the time constraints of a given class period. We encourage other instructors to consider implementing activities that create opportunities to introduce a topic through student exploration outside of the classroom.

Further, group exploration activities that are used to introduce a topic are typically designed with the teacher's role in mind; that is, consideration is given to what ideas students might come across and how the teacher might scaffold those ideas for class discussion. With these activities occurring in the beginning of class, usually with an entire lesson to be taught afterwards, timing is also often a major consideration. If one group comes across an important idea in their exploration, the teacher may take this opportunity to bring the discovery to everyone's attention to streamline the process and ensure the warm-up is completed before moving on. While this can lead to meaningful group discussions, this can also result in a scenario where the groups that worked fastest tend to have their ideas broadcast to the class, while the groups that worked slower may not have a chance to share their insights. Even though there are group work strategies that seek to mitigate the influence of dominant individuals within a group (e.g. Cohen & Lotan, 2014; Featherstone et al., 2011; Horn, 2012), it can be more difficult to prevent a dominant group from taking over the classroom discussion. Not only did the online implementation allow each student to individually develop their ideas before the group discussion, but it also reduced the influences of the instructor and other students on the development and sharing of those ideas. As such, we recommend that instructors (1) design open-ended tasks to introduce a topic, (2) provide questions to guide students' reflection on the tasks, and (3) have students engage with the task and reflection at home before the next class meeting. These steps create opportunities for students to explore and make hypotheses about course content before it is discussed in class.

Another key takeaway from the Online Measurement Activity is that allowing for students to share their work in an online environment can lead to more creative approaches and presentations, as well as opportunities for students to share in-progress thinking before whole class discussions. We found that collecting all student work in one document made it easier to present a variety of student work and created an opportunity to elevate a wide range of student perspectives. A shared class workspace, such as Google Slides, allows every student to have access to other students' work while engaging with the activity at home. We encourage instructors to reflect on ways to create shared workspaces for students to collaborate and to learn from and with one another outside of the classroom environment. Specifically, we recommend that instructors (1) give students creative freedom and flexibility to decide how to share their work, (2) incorporate shared workspaces (such as Google Slides) for students to collaborate both in class and outside of class, and (3) create opportunities for students to present their work using these shared workspaces.

By redesigning this exploratory activity to be implemented at home, we are able to reflect on the constraints we may be imposing (intentionally or not) on students when we implement these activities in the classroom. These constraints can be literal (e.g. materials provided, time allowed) or they can be more subtle, such as the assumptions we make as instructors about our role in the activity. While intentionally sequencing student ideas plays a very important role in shaping class discussions, holding back and allowing students to fully express their ideas with less guidance can also yield valuable insights and discussions. Moreover, allowing students to work on these explorations individually and with no time limit meant that everyone had an opportunity to contribute to the discussion.

Disclosure statement

No potential conflict of interest was reported by the authors.

ORCID

Jessica Gehrtz http://orcid.org/0000-0002-3564-6456
Raquel Vallines Mira http://orcid.org/0000-0003-2396-4948
Priya V. Prasad http://orcid.org/0000-0003-2025-6727

References

Boaler, J. (2015). *Mathematical mindsets: Unleashing students' potential through creative math, inspiring messages and innovative teaching*. John Wiley & Sons.

Cohen, E. G., & Lotan, R. A. (2014). *Designing groupwork: Strategies for the heterogeneous classroom, a new edition* (3rd ed.). Teachers College Press.

Esmonde, I. (2009). Ideas and identities: Supporting equity in cooperative mathematics learning. *Review of Educational Research*, *79*(2), 1008–1043. https://doi.org/10.3102/0034654309332562

Featherstone, H., Crespo, S., Jilk, L. M., Oslund, J. A., Parks, A. N., & Wood, M. B. (2011). *Smarter together! Collaboration and equity in the elementary math classroom*. National Council of Teachers of Mathematics.

Horn, I. S. (2012). *Strength in numbers: Collaborative learning in secondary mathematics*. National Council of Teachers of Mathematics.

Jansen, A. (2020). Rough-draft thinking and revising in mathematics. *Mathematics Teacher: Learning and Teaching PK-12*, *113*(12), e107–e110. https://doi.org/10.5951/MTLT.2020.0220

Lotan, R. A. (2003). Group-worthy tasks. *Educational Leadership*, *60*(6), 72–75.

OPEN ACCESS

Peer-Assisted Learning in a pandemic

Jessica Hargreaves, Claire Ketnor, Ellen Marshall and Sue Russell

ABSTRACT

Peer-Assisted Learning (PAL) schemes typically involve student volunteers (PAL Leaders) designing and delivering sessions that support groups of students in lower years with their studies. This paper discusses three different PAL schemes, within Mathematics degrees at Sheffield Hallam University (a Post-92 University) and The University of York (a Russell Group University), which took place entirely remotely during the Covid-19 pandemic. In this paper, we explore the challenges and benefits of remote PAL schemes via the three case studies, from both staff and student perspectives. There are aspects of PAL in a pandemic that we would (or would not) take forward to a blended or in-person approach in the future. In particular, we found that, while students and staff alike are looking forward to in-person contact to provide them with interactions and easier methods of communication, the remote provision for peer support has real value in terms of accessibility and inclusivity. Furthermore, integrating technology enhanced learning into in-person sessions can provide added value.

1. Introduction

Peer-Assisted Learning (PAL) schemes in the UK are typically based on the SI-PASS (Supplemental Instruction Peer-Assisted Study Sessions) model, with trained PAL Leaders working collaboratively with staff to provide students with a peer learning environment (see, Arendale, 1994, for example). There is variation between schemes, but some of the core principles of PAL are that it is designed to be run by students, for students and that it involves working with 'students as partners' (Sedghi & Lunt, 2015). The PAL sessions are usually run by trained students, in later years of study, for first-year students. Research in the UK (Dawson et al., 2014; Keenan, 2014) shows that PAL schemes not only enhance transition into university, but can also improve student progression, achievement and satisfaction.

During the 2020–2021 academic year, UK COVID-19 restrictions required many aspects of university courses to be offered remotely (Nerantzi, 2020). This transition resulted in many educators adapting their traditional classroom-based teaching approaches and materials to the online space (Crawford et al., 2020). On-campus PAL sessions are typically characterized by participation in interactive and collaborative activities (Sampaio et al., 2011). Therefore, for PAL coordinators, the goal was to devise alternative remote PAL provision to provide their learners with an equivalent PAL experience.

Within this classroom note, we present details of three different PAL schemes that take place within Mathematics degrees and were all run remotely during the 2020–2021 academic year. This article is organized as follows. In Section 2, we introduce the three case studies of PAL schemes, highlighting their similarities and differences. In Section 3, we explore the challenges and benefits of remote PAL schemes via these case studies, from both staff and student perspectives. Section 4 concludes by outlining the 'takeaways' from delivering PAL in a pandemic, which are the approaches and principles that will be of use in future post-pandemic practice.

2. PAL case studies

This section starts with an overview of the three PAL schemes, including their similarities and differences, before outlining the details of remote delivery in Section 2.2.

2.1. Overview

The PAL scheme in Case Study 1 was integrated into the Mathematics course at Sheffield Hallam University (the Post-92 University) in 2008–2009 with the main aims of easing the transition into university and fostering a learning community (Waldock, 2011). Students in the first semester of their degree undertake a summative assessment task in groups, whilst being supported by PAL Leaders during timetabled sessions. These 'PAL projects' draw on skills developed across the course but are run independently from other modules.

Unlike the PAL scheme in Case Study 1, an alternative approach to peer-assisted learning is to follow the 'supplemental instruction' model and embed the PAL sessions within the teaching provision of a specific 'historically difficult' module (Arendale, 1994). In this context, 'difficult' implies that students find it a challenge to understand the material covered during lectures and identify that they require additional support beyond lectures and seminars. Statistics is a subject that students often have a negative perception of, or feel anxious about, which impacts on their ability to learn (Cornock et al., 2019; Marshall, 2019). Therefore, in 2019–2020, peer support was introduced for the first year 'Probability and Statistics' module at Sheffield Hallam University (Case Study 2). Interestingly, when mathematics students from the University of York (Case Study 3) were consulted on which module they felt would benefit most from PAL provision in 2020–2021, they identified the equivalent 'Probability and Statistics' module in their degree.

The PAL sessions in Case Study 2 operate mainly as 'drop-in' sessions, directed by the needs of those that attend, giving students the opportunity to: get support with understanding the content of the module (including developing skills such as SAS programming and report writing) and (from 2020 to 2021) seek guidance on the coursework. However, the PAL sessions in Case Studies 1 and 3 include activities that are planned by PAL Leaders in advance. For example, PAL Sessions in Case Study 3 have the following structure:

(1) **Drop In- 'ask us anything':**
This was intended to ease the transition and help students settle into university life (Byl et al., 2015).
(2) **'Legal Hints'- tips for the weekly assessment (approved by lecturer):**
Providing this information was mainly to improve attendance, so students see a tangible benefit to attending (Sloan et al., 2020).
(3) **Interactive pre-planned activity**
PAL Leaders facilitate activities to encourage discussion amongst the group, to develop comprehension of previously taught module content (Sampaio et al., 2011).

In all three case studies, PAL Leaders have (generic) SI-PASS equivalent training from qualified instructors plus additional subject-specific training from staff in the course teams. The training includes exploration of scenarios that could be encountered (e.g. if PAL Leaders do not know the answer or challenging group dynamics). PAL Leaders are also provided with techniques and tools to encourage engagement as they facilitate PAL sessions. PAL Leaders for the statistics PAL schemes (Case Studies 2 and 3) had a specific session which was adapted from Sigma training for mathematics and statistics support tutors (Owen et al., 2015). Furthermore, in all three Case Studies, the PAL Leader training sessions adopt the techniques that the PAL Leaders will utilize in the sessions- for demonstration purposes (see Section 2.2.2).

All three schemes were designed so that there was no direct staff presence at the PAL sessions. However, staff provide background support for the PAL Leaders- including: the initial training; support sessions and regular communication on how the sessions are going. In Case Study 3, PAL Leaders and staff attended a weekly 'de-brief' session. The de-brief session gave the PAL Leaders the opportunity to feedback and reflect on the previous sessions and then work collaboratively with staff to plan the next session. The collaborative nature of this session emphasized the central ethos of PAL, which is the principle of working with the students as partners throughout the scheme (Sedghi & Lunt, 2015).

A summary of the three schemes is contained in Table 1.

2.2. Remote delivery of the PAL schemes

In this section, we present details of the remote delivery of the schemes during 2020–2021. Pre-pandemic, the PAL schemes at Sheffield Hallam University were already successful at enhancing feelings of belonging and helping students transition into university (Cornock, 2016). Staff for all three schemes were keen to replicate this experience as much as possible in the remote environment.

2.2.1. Remote set-up

All three PAL schemes took place on Zoom. In traditional PAL sessions, staff tend not to be present. However, when delivered remotely, staff at Sheffield Hallam University remained in the Zoom room in some capacity, initially more as a 'receptionist' but also to provide assistance and support. For Case Studies 1 and 2, groups used the breakout rooms (either for meetings or for drop-in support). In Case Study 1, the PAL Leaders stayed with their groups and the PAL Supervisors moved between groups and the main Zoom room. In Case Study 2, first-year students could use the breakout rooms to get drop-in support or

Table 1. A summary of the similarities and differences between the three PAL schemes.

Institution	Case study 1 (Sheffield Hallam University)	Case study 2 (Sheffield Hallam University)	Case study 3 (University of York)
First established	2008–2009	2019–2020	2020–2021
Embedded within a module	Contributes to a Mathematics module mark, but run independently	First (and later second) year 'Probability and Statistics'	First year 'Probability and Statistics'
Session type	PAL Leaders assigned to specific project groups and plan support sessions.	Group or individual 'drop-in' support	Structured sessions (planned with staff)
PAL supervisors (students who support PAL sleaders)	Yes	Yes	No
Staff involvement	Support in the background	Support in the background	Planning and debrief

Table 2. Google Doc activity in the training for PAL Leaders and supervisors (Case Study 1).

Topic	Who	Findings (including links to websites)
Wait time	PAL leaders	
Peer interaction	PAL leaders	
Starting a meeting	PAL leaders	
Closing a meeting	PAL leaders	
Peer observation	PAL supervisors	

could work on their group assessment task. Support would be on hand from any of the PAL Leaders. Case Study 3 did not have a staff presence at the PAL sessions.

2.2.2. Remote training and PAL Leader tools

As all PAL sessions would be delivered remotely during 2020–2021, the training sessions replicated this modality. This gave PAL Leaders vital experience in hosting remote sessions, whilst simultaneously giving them the experience of participating in them.

Note, in contrast to Case Studies 1 and 2, the PAL scheme in Case Study 3 was first introduced during the pandemic and therefore was designed to run remotely from inception. Therefore, the interactive pre-planned activities within the PAL sessions (see Section 2.1) would be delivered remotely and hence utilize online tools such as Jamboard, Padlet, Kahoot and Zoom features (e.g. chat function, emoji reactions, polls, break-out rooms, etc.). Therefore, the PAL Leader training also utilized and introduced these tools.

However, most of the methods provided in the traditional training for (the established) Case Studies 1 and 2 involved physical activities to encourage active learning (e.g. the use of post-it notes and individual whiteboards). Therefore, in 2020–2021, alternative methods were provided to the PAL Leaders to encourage engagement within the sessions. These included: online electronic games as ice-breakers, instead of the physical 'Maths Arcade' (Cornock, 2015); virtual post-it notes; the use of the virtual whiteboard (with ticks and crosses on the boards for voting), the Zoom private chat function and Socrative (2021), instead of using individual whiteboards; and Google Docs, for collecting information in groups. An example of a table which formed an exercise in the PAL training from Case Study 1 is presented in Table 2. The students had to find information and resources and fill in their part of the table within Google Docs within 10 minutes.

3. Reflections on the remote delivery of PAL

In this section, we reflect on the range of experiences delivering PAL remotely in a pandemic from both staff and student (attendees, PAL Leaders/ Supervisors) perspectives.

3.1. Methods

In order to understand the student perspective, authors at Sheffield Hallam University interviewed PAL Supervisors from both schemes who had delivered peer support face to face in 2019–2020 and remotely in 2020–2021. At the University of York, PAL Leaders and potential attendees (i.e. students enrolled on the module) were invited to give feedback on the scheme through a short questionnaire (which had 15 respondents).

3.2. Overall PAL experience

Although, overall, delivering PAL remotely was more challenging, and both PAL supervisors at (Sheffield Hallam Univeristy) concluded that the scheme was better in person, the key aims (see Section 1) of PAL were achieved. They thought that running the PAL schemes remotely did still help form a community and that it would have been particularly difficult for students to have made friendships with other people on the course in 2020–2021 without it. This aspect was also identified by the students at the University of York with one student commenting 'I got to meet a lot of new people in the mathematics department, very valuable especially this year'.

All those surveyed at the University of York felt that they had benefitted from being part of the PAL scheme. For example, all (15) students surveyed (including PAL Leaders) felt being involved with PAL had improved their subject knowledge.

3.3. Challenges and benefits of remote delivery

The main benefits of remote peer support centred on the convenience and accessibility of remote sessions compared to on campus. For staff, not needing to book a physical space for the session or restrict group sizes and not requiring anyone to come onto campus especially for a session. Students also commented on this aspect ('You don't feel like you're wasting anybody [sic] time - it's easier and more convenient... but when it's in person it makes it less convenient and makes you wonder if it's worth it going all that way for just a small question'.), as well as feeling more anonymous '[Online] can be a little less nerve wracking because you are on the other side of the screen, you can be anonymous'.

Attendance for Case Study 1 remained consistent with previous years but for Case Study 2, attendance increased. This could partly be due to the convenience of online sessions but the introduction of direct support with group work (in line with Case Study 1) is also likely to have impacted on engagement. One of the main challenges to delivering peer support remotely was using and managing the virtual space. In a traditional on-campus peer support session, PAL Leaders would be aware of new students entering the room, or which students are requiring help, and move easily from table to table without disruption, whereas this was harder online.

PAL Supervisors also commented on the negative impact of remote delivery on communication, between the students themselves and with PAL Leaders ('The general conversation seemed different as there was often minimal response and seemed like a chore to get some people communicating.'; 'I think some people generally have more anxiety when joining Zoom rooms and speaking, it always seemed like the routine was to join and sit in silence until spoken to'.). They found it more challenging to get students to participate as it is easier to avoid joining in remotely, and this may have 'limited the friendships that could have been made'. They acknowledged that this became easier after the initial reluctance. They emphasized the importance of the scheme in helping students form support groups and having a relaxed atmosphere to enable this to happen, which can be done more easily in-person.

Survey results from the University of York suggested a strong preference for having both online and in-person options available (73%) with both individual options also receiving support (online only (7%) and in-person only (20%)). These findings suggest that students would like to retain a remote option in future.

4. Take-aways

Following reflection, we are able to report a number of take-away messages. In Section 4.1, we outline generic tips for delivering PAL schemes. In Section 4.2, we outline how the experience of delivering remote PAL schemes informs future practice.

4.1. Generic tips for PAL schemes

As our three examples illustrate, there is no single way to run a PAL scheme. However, we have some general considerations.

Carefully linking a PAL scheme to assessment, such as providing targeted support for coursework, seems to improve attendance.

The amount and type of staff involvement needs to be considered. We would recommend keeping direct staff involvement at PAL sessions to a minimum, whilst still maintaining high levels of support for students running the scheme.

Activity-based training is not only an effective way to convey information, but also provides PAL Leaders with techniques and tools that they can adapt for their own PAL sessions, to encourage engagement. Having a variety of techniques gives PAL Leaders a number of approaches they can use to involve all students.

4.2. Remote PAL schemes informing future practice

To summarize, these case studies have illustrated that the main benefits of remote PAL sessions are the flexibility (Vlachopoulos & Jan, 2020) and option for anonymity afforded by this delivery method. However, although staff and student feedback indicated that the remote PAL sessions had, despite the challenging circumstances, developed a sense of community and enabled interactive activities, most participants felt that, for these aspects, the in-person experience is preferred. Therefore, Case Studies 1 and 3 (where the sessions include pre-planned interactive activities) plan to provide face-to-face PAL Sessions in future (when possible).

However, the intrinsic structure of a 'Drop-in' or 'Question and Answer' session facilitates the transition to a remote setting, as most aspects of the in-person experience are retained, with the additional benefits outlined previously. Therefore, in two of the case studies (those embedded within modules), we have identified that some remote provision will be retained in future. In particular, in Case Study 2, 'virtual group work rooms' (for groups to meet with support on hand at specific times) will be retained. In Case Study 3, there will also be a dedicated Online PAL Drop In Session. This provision will ensure that the PAL schemes are more flexible for students, preserve anonymity (if required) and, hence, are more inclusive. Furthermore, the dual option of in-person and remote delivery will hopefully mean that more students engage with the sessions (than if there is only one option).

Furthermore, the experience of teaching and learning remotely has exposed both staff and students to a range of online tools that can also be utilized in traditional settings. Using anonymous methods of communication in in-person sessions can be beneficial, as an efficient method to gauge understanding and to encourage participation. Therefore, all three schemes will embed technology enhanced learning in future.

Disclosure statement

No potential conflict of interest was reported by the authors.

ORCID

Jessica Hargreaves http://orcid.org/0000-0002-7173-7902

References

Arendale, D. (1994). Understanding the supplemental instruction model. In D. C. Martin & D. Arendale (Eds.), *Supplemental instruction: Improving first-year student success in highrisk courses* (2nd ed., pp. 3–10). National Resource Center for the Fresh- man Year Experience and Students in Transition.

Byl, E., Struyven, K., Meurs, P., Abelshausen, B., Lombaerts, K., Engels, N., & Vanwing, T. (2015). The potential of peer assisted learning as a tool for facilitating social and academic integration. *Journal of Learning Development in Higher Education*, (Special edition: Academic Peer Learning), November, 1–28. https://doi.org/10.47408/jldhe.v0i0.343

Cornock, C. (2015). Maths Arcade at Sheffield Hallam University: Developments made in a new space, *MSOR Connections, 14* (1). https://doi.org/10.29311/ndtps.v0i14.3270

Cornock, C. (2016). The evaluation of an undergraduate Peer Assisted Learning Scheme at Sheffield Hallam University, *Journal of Learning Development in Higher Education (Special Edition - Academic Peer Learning, Part II).*

Cornock, C., Marshall, E., & Hargreaves, J. (2019). Work-related learning: Benefits and challenges. *New Directions in the Teaching of Physical Sciences, 14*(1). https://doi.org/10.29311/ndtps.v0i14.3270

Crawford, J., Butler-Henderson, K., Rudolph, J., Malkawi, B., Glowatz, M., Burton, R., Magni, P.A., & Lam, S. (2020). COVID-19: 20 countries' higher education intra-period digital pedagogy responses. *Journal of Applied Learning & Teaching, 3*(1), 9–28. https://doi.org/10.37074/jalt.2020.3.1.7

Dawson, P., van der Meer, J., Skalicky, J., & Cowley, K. (2014). On the effectiveness of supplemental instruction: A systematic review of supplemental instruction and peer-assisted study sessions literature between 2001 and 2010. *Review of Educational Research, 84*(4), 609–639. https://doi.org/10.3102/0034654314540007

Keenan, C. (2014). *Mapping student-led peer learning in the UK.* Higher Education Academy. Retrieved May 28, 2021, from https://www.heacademy.ac.uk/system/files/resources/Peer_led_learning_Keenan_Nov_14-final.pdf.

Marshall, E. (2019). Embedding and assessing project based statistics. *MSOR Connections, 17*(2), 75–82. https://doi.org/10.21100/msor.v17i2.979

Nerantzi, C. (2020). The use of peer instruction and flipped learning to support flexible blended learning during and after the development and implementation of a Peer Assisted Learning programme at the COVID-19 pandemic. *International Journal of Management and Applied Research, 7*(2), 184–195. https://doi.org/10.18646/2056.72.20-013

Owen, A., Marshall, E., & Smith, S. (2015). Scenario based training of statistics support tutors. In *IMA Barriers and Enablers conference proceedings 2015.*

Sampaio, P. N., Teixeira, J. M., Camacho, M. F., & de Freitas Gouveia, R. H. (2011). Blended peer-assisted learning platform: Improving learning outcomes with a collaborative environment. *Journal of Educational Technology Systems*, *39*(4), 371–395. https://doi.org/10.2190/ET.39.4.c

Sedghi, G., & Lunt, T. (2015). The development and implementation of a Peer Assisted Learning programme at the University of Liverpool. *Journal of Learning Development in Higher Education, Special Edition: Academic Peer Learning*, November 2015, 1–17.

Socrative. (2021). Showbie Inc. Retrieved May 28, 2021 from https://www.socrative.com/.

Sloan, D., Manns, H., Mellor, A., & Jeffries, M. (2020). Factors influencing student non-attendance at formal teaching sessions. *Studies in Higher Education*, *45*(11), 2203–2216. https://doi.org/10.1080/03075079.2019.1599849

Vlachopoulos, P., & Jan, S. (2020). Exploring modes of lecturing as a teaching method in Higher Education: Student attendance, preference and motivation. *Journal of University Teaching & Crawford, J. Learning Practice*, *2020*(17), 1–18. https://doi.org/10.53761/1.17.5.15

Waldock, J. (2011). Peer assisted learning. In Waldock, J. (ed.) Developing graduate skills in HE mathematics programmes. national HE STEM programme (pp. 22–23) [Online]. Available at: http://www.mathcentre.ac.uk/resources/uploaded/gradskills.pdf

Checkpoints for online mathematics students with learning challenges

James William Dalitz

ABSTRACT

Many mathematics students require support for learning challenges, such as those invoked by mathematics anxiety. Support for these learning challenges is imparted through personal interactions between the student and the instructor, support staff or their peers. The online delivery of mathematics classes and support meetings reduced the quality of the interactions compared to face-to-face interactions. This article presents an alternative approach used to support students with learning challenges in a virtual learning environment. The approach reported was used in a four week online preparatory mathematics subject. Most of the students were mature age, some years out of formal mathematics education, and low in confidence. In preparing the modules the nature of the subject and the cohort was considered and additional content called Checkpoints was added. The Checkpoint pages were designed to alleviate the learning challenges the students faced by including signposting, supporting positive dispositions, providing alternative learning strategies, and sharing assessment strategies. Feedback showed students engaged with the Checkpoints and found them helpful, leading the students to reflect on their learning. The checkpoints also 'chunked' the math content, resulting in a 'nice break', and provided students with reassurance they were on-track.

1. Introduction

The shift to online learning as a result of the COVID-19 pandemic has further complicated the existing challenges students face with studies involving mathematics or numeracy (Lambie & Law, 2020). First-year university students studying courses including Science, Engineering, Education, Nursing, Computing and Economics are required to undertake either mathematics subjects or subjects that have mathematics embedded in the form of numeracy elements (Khasawneh et al., 2021; Taylor & Galligan, 2006). Research has consistently shown that many students experience challenges with the mathematical demands of these subjects (Khasawneh et al., 2021). These challenges can often be attributed to a combination of under-developed academic skills (Rylands & Shearman, 2018), mathematics anxiety and negative mathematical dispositions (Jackson & Leffingwell, 1999; Perry, 2004).

Support for students with mathematical learning challenges pre COVID-19 was primarily provided through face-to-face interactions with academic staff (Exley & Dennick, 2004), support staff (Blazer, 2011) and fellow students (Geist, 2010; Harper & Daane, 1998; Woodard, 2004). However, following the shift to online learning, the lack of visible and audible cues has diminished the quality of these interactions (Lambie & Law, 2020). This raises questions regarding the support of students with learning challenges in a virtual environment. This article presents a proactive approach called Checkpoints used in a virtual learning environment to support students experiencing challenges with their mathematics learning.

2. Background

First-year university students typically face challenges related to the transition to independent learning and independent living (Kift, 2015). Students are often confused by incomplete perceptions and limited expectations of university life and overwhelmed when these combine with health, financial and life balance stressors. Universities typically counter these challenges by operating sponsored actions at an institutional and faculty level and a variety of peer-based networks sponsored by student associations.

The shift to online learning in response to the COVID-19 pandemic further complicated the challenges of transitioning to first-year students. This section explores the challenges brought about by the rapid transition to online learning and the implications for some first-year university students studying mathematics.

2.1. Studying mathematics at university

Studying subjects with mathematics or numeracy at university has proved a challenge for Australian students (Taylor & Galligan, 2006) as students are often underprepared academically (Rylands & Shearman, 2018) and/or suffer from mathematics-induced anxiety (Perry, 2004; Taylor & Galligan, 2006). The normal support mechanism for mathematics anxious students is face-to-face interactions with their peers and support staff (Harper & Daane, 1998) with some students prefer one-to-one contact as the most effective method for addressing mathematics anxiety (Núñez-Peña et al., 2015).

Many universities have specific mathematics help centres or learning advisers that support students who are underprepared by offering mathematical support aimed at improving performance and retention (MacGillivray & Croft, 2011; Rylands & Shearman, 2018). These centres primarily operated in a face-to-face manner with staff focusing on providing positive environments that are conducive to learning for students with mathematics anxiety and other mathematics learning challenges (Woodard, 2004).

The transition to online learning prompted by the COVID-19 protocols limited the opportunities for students to connect with academic staff, support staff and their peers. While online meetings were still possible, the reduced quality of these interactions lessened the effectiveness of these meetings (Lambie & Law, 2020). The Checkpoints described in this paper are online pages comprising information that is contextually relevant but specifically developed to mimic the support normally provided by the academic staff, support staff and peers.

2.2. *Supporting learners of mathematics*

Students who are underprepared and anxious about doing mathematics have struggles with starting their mathematics study (Mann et al., 2016), and then having the resilience to maintain their study (Marshall et al., 2018). Foremost, these students require regular positive support to encourage them to persist with their studies and maintain their intensity (Marshall et al., 2018). Further supports for students with mathematics learning challenges can include a variety of strategies: structuring their learning to provide breaks; sharing strategies that help students manage their feelings; and providing different ways of understanding the content.

Inserting Checkpoints at regular intervals provides the students with a structure to plan their study by breaking the mathematics content into manageable chunks. Regular breaks reduce the build-up in cognitive load (Young et al., 2012), and consequentially the build-up of anxiety (Alamolhodaei, 2009), thereby increasing the students' chances for success. Small successes build up a student's self-efficacy and counteract their anxiety (Marshall et al., 2018).

Impending assessment can exacerbate many students' feelings of unease and have consequent impacts on their performance (Perry, 2004). Having students practise assessment tasks and engage in alternative approaches to completing tests can reduce their anxiety (Mann et al., 2016). Similarly, encouraging students to practise using alternative learning strategies at home can help desensitize them to mathematics anxiety (Sheffield & Hunt, 2006).

Providing alternative ways of thinking mathematically gives students extra strategies for learning mathematics that could support their understanding (McGlaughlin et al., 2005). A model used in mathematics education subjects to assist pre-service teaching students to develop a broader understanding of mathematical concepts describes four different ways of thinking mathematically: essence; contexts; connections and communication. Traditional mathematics teaching focuses primarily on essence, or the teaching of procedural skills, and has been linked to causing mathematics anxiety (Finlayson, 2014). Cornell (1999) noted the learning gaps lead to anxiety. Developing the students' capacity for the three other ways of mathematical thinking by focusing on communication, contexts and the connections can assist students to fill their learning gaps. The Checkpoints provide content that supports diversity in the students' thinking and approaches to learning while also giving the subject structure and fostering positive dispositions.

3. Teaching context

The Checkpoints discussed in this paper were embedded in a four-week intensive mathematics subject. The students enrolled were attempting the mathematics subject only. The subject comprised three online modules, each comprising up to 10 topic pages and 3–4 Checkpoints. The intensive offering provided 20 h of contact, as compared to 39 h for the same 13-week subject. The subject is part of the program that provides access to tertiary studies for students who did not meet the minimum knowledge requirements. The access programs focus on enhancing the academic preparedness of the students by improving their content knowledge.

The mathematics subject contained a diverse range of learners. The students self-identified as needing to improve their mathematics, either through self-reflection following a diagnostic test or in response to the levels of assumed knowledge of their intended tertiary courses. For most of the students, this was their first return to education in many years and their first tertiary subject. These two factors raised staff concerns about the student's ability to be successful at learning mathematics in an online subject.

The redesign of the subject for online delivery due to COVID-19 was an opportunity to integrate Checkpoints as supports to address the staff concerns about the learners. The redesign involved transitioning the preparatory mathematics subject resources from traditional static documents, such as Word documents and PDFs, to interactive resources using the H5P platform ready for online learning. The H5P pages containing mathematics content were labelled as Topics and the non-mathematical content pages were labelled as Checkpoints.

4. Checkpoints

The Checkpoints aimed to fill the gap left by the reduction in support typically provided by academic and support staff and peers. This followed the recognition that many of the traditional support mechanisms such as help centres, learning advisers and staff and peer interactions were somewhat curtailed by the pandemic protocols. Noting the online delivery would affect personal interactions, the Checkpoints focused on providing support for the students' wellbeing and learning. The Checkpoints were also 'tailored' (Mann & Willans, 2020, p. 56) with content that would support the students' online learning by supporting them to take ownership of their learning and control how they learn in terms of timing, contexts and strategies.

Checkpoints were developed by integrating combinations of different elements that embodied strategies to counter the learning challenges the students face learning mathematics online. These elements focused on signposting upcoming content, supporting wellbeing, sharing learning tips that highlight alternative ways of thinking mathematically, and providing strategies for assessment preparation.

4.1. Signposting

Signposting is used to inform the students about the content by connecting it to prior and upcoming content, and to support engagement by providing relevant contextual information. The excerpt below (Figure 1) explicitly links the decimal fraction content to prior content (powers of 10) and upcoming learning (decimals), which assists students to make connections between mathematical concepts and ideas, and informs the students of some contexts (money, percentages) where decimal fractions are found.

4.2. Wellbeing

The checkpoints support the students' disposition, motivation and willingness by providing regular breaks. In these breaks the content shifts to focus on promoting positive thinking, providing alternative strategies for learning and preparing for assessment, and

Checkpoint 1.4: Learning and support strategies

In the next few topics the focus is on developing an understanding of fractions with a focus on decimals fractions and decimal numbers.

The increased usage of mathematics in society has had to evolve to include the possibility of part values or fractions of a whole. This gave rise to the values we know as fractions such as a half or a quarter. Decimals fractions are a special type of fraction where the denominator is 10 or a power of 10 (100, 1000, and so on). Decimal fractions (e.g, 4/10) and the equivalent decimals (0.4) are important mathematics concepts as they occur regularly in everyday life when we consider contexts such as money, weight or percentages. Several resources are provided to give students opportunities to develop an understanding of fractions with a focus on decimals and decimal fractions.

Figure 1. Signposting in a Checkpoint providing context and connections for upcoming content (Dalitz, 2021a).

In fact, one of the best strategies for getting better at maths is using your maths skills and knowledge as much as you can outside your study times. For topics like percentages, rates and ratios there are many opportunities to practice including that has using percentages when shopping at sales, rates such as your speed to guestimate your arrival time or unit prices when grocery shopping, i.e., price per kg, 3 for $$$, using ratios when cooking or sharing. Make the most of these opportunities to practice your maths and your learning will benefit.

Figure 2. Using positive motivational tips to provide advice about normalizing mathematics (Dalitz, 2021b).

supporting students to reflect on their progress to date. An important element in supporting the students learning is to modify the student's relationship with mathematics by addressing the mathematics anxiety or negative dispositions as these can lead to mathematics avoidance or the delaying of attending to mathematics tasks such as study and assessments (Woodard, 2004).

Using mathematics in everyday situations that are low stakes like shopping, as suggested in Figure 2, can normalize mathematics and reduce anxiety when the student needs to do mathematics work or assessments (Mann et al., 2016).

Students suffering from mathematics anxiety are less likely to seek support (Warwick, 2008). Symonds et al. (2008) suggest using a proactive approach that directs the student to support. The third paragraph in Figure 3 provides the students with directions to the institutionally provided support (Figure 4). This emphasizes that students are not alone and advises where they can seek help.

4.3. Learning tips

To support the students' learning, regular references were made to the mathematical contexts, communication and connections. These references provide students with additional ways of thinking about the specific mathematics concepts and were intended to help them to understand the concepts. In Figure 5 using the real-life context of money, and specifically coins can support the students understanding of decimal value and operations (Mann et al., 2016). Providing scaffolding (Figure 6) can help students make sense of the various parts of the unit names used to communicate in measurement. Figure 7 shows a basic progression of mathematical operations, showing how the algorithms rely upon mastery

Positive Motivational Tip

The good news is that the subject is coming to an end and you'll soon be ready to move onto the next stage of your learning, and the bad news is probably also that the subject is coming to an end as this means exams for MA0022. If this is concerning you should know that there is still time remaining, and there is plenty of help available, to improve your maths and reduce your concerns. You can still make a big difference to your maths knowledge and skills in the time remaining before the exam. To achieve effective improvement will require you putting in work and being strategic. This may mean making short term sacrifices and briefly changing your practices.

First, regarding putting in the work, take stock of how you are going and how much maths you're doing each week. If you're not doing as well as you'd like then assess whether you're putting in your 30 hours of study per week. If you are doing 30 hours per week but still not getting the results you want, you might need to consider increasing this time for the week or so remaining, especially if MA0022 is a pre-requisite for the next stage of your study.

Secondly, being strategic relates to knowing what to do to produce real improvement and focusing on these areas or actions. However, you might not be sure of what areas or actions you should focus on. The Learning Centre is a fully funded free unit operating within the Division of Student Life. As a JCU student you can make use of the free resources, of which many relate to mathematics, and also request consultations with the Learning Advisers. The Learning Advisers are professional educators who can help with your learning experience including planning your study, preparing for exams assisting with specific assessments.

Figure 3. Using positive motivational tips to highlight support mechanisms available for the students (Dalitz, 2021c).

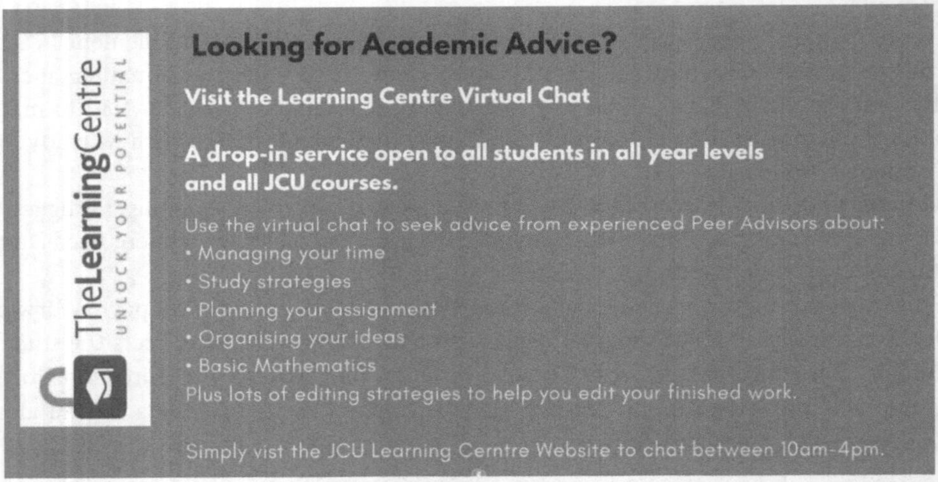

The Learning Center website ... https://www.jcu.edu.au/students/learning-centre

Maths and Stats resources ... https://www.jcu.edu.au/students/learning-centre/maths-and-statistics

To book a Learning Adviser consultation ... https://www.jcu.edu.au/students/learning-centre/contacts/contact-form

Figure 4. Details for contacting the institutional student support services (Dalitz, 2021c).

of earlier operations and basic mathematics facts. These alternative ways of thinking can support students to develop their own strategies for understanding mathematics (Mann et al., 2016).

Learning tip

One way of learning mathematics is to relate the mathematics to familiar contexts. For example, you can relate decimal calculations to common decimal contexts like money or measurements to help you make sense of the calculations and determine the reasonableness of results (that is, whether the answer seems correct using logic).

For example, is 13 x 0.05 more or less than 1? If I change my thinking to 1 as $1 and 0.05 as 5 cents, then I know I need 20 x 5 cents for a dollar so 13 x 5 cents will be less than a dollar. I could then add up 13 x 5 cents to get my answer in cents and then convert it into dollar to get the final answer … 13 x 5c = 65c = $0.65.

Figure 5. Suggesting coins as a context for learning about decimals (Dalitz, 2021a).

Learning tip

Communication, the language of measurement, broad definitions not memorisation of formula.

One way of learning mathematics is to focus on the language and ways of communicating.

Knowing the key units for each type of measurement and the significance of the prefixes makes it easy to work with different units and convert between units. Similarly, knowing how to use scientific notation makes calculations with very large or very small numbers easier.

Spend some time getting to know the language of measurement and comprehending and learning will be easier for you.

Figure 6. Suggesting that measurement is easier to learn if you focus on understanding the naming of units (Dalitz, 2021a).

Learning tip

One way of improving your understanding and retention of mathematics is to focus on the connections between different topics and concepts.

Some of the connections you may have noticed over the last few weeks include:

1. The addition and subtraction algorithms are both based on place value.
2. The multiplication algorithm also uses addition.
3. The division algorithm uses multiplication and subtraction.
4. The four algorithms have been extended with a few extra rules to include integers.
5. The algorithms were also extended using a few changes to include decimals.
6. Measurement often involves decimals.
7. Converting units uses multiplication and division and can involve decimals

The knowledge of algorithms is used to calculate measurement solutions using formulas.

The diagram below show the heirarchy of knowledge for this module. To be good at algorithms you need good number sense and mental maths. Strong algorithms will make integers, decimals and unit conversions easier and in turn, this makes measurement easier.

Figure 7. Highlighting connections between different operations and different types of numbers (Dalitz, 2021a).

Test study tip

Successful study means using your time effectively to improve your knowledge and your grade. This means knowing what to study to improve your knowledge and starts with preparation for study. You should organise your subject materials and focus on identifying what you know and what you need to study. Use the following list to start planning your study.

Figure 8. Prompting students to plan for upcoming assessment (Dalitz, 2021b).

Test study tip

Successful study means using your time effectively to improve your knowledge and your grade. From your experience with Test 1, you should have some idea as to how effective your study strategies are. Think back on how Test 1 went and what you might do the same and what you could do differently that can improve your result.

Figure 9. Reflecting on test preparation strategies (Dalitz, 2021a).

4.4. Assessment tips

Assessment is a critical part of tertiary education and causes significant issues for many students with mathematics anxiety (Young et al., 2012). The increased anxiety induced by mathematics assessment increases the student's neural loading, decreasing their capacity to retain and apply their knowledge (Ashcraft, 2002; Ashcraft & Kirk, 2001; Young et al., 2012). To manage their anxiety, the students need to become better learners and acquire better test preparation techniques.

Assessment tips help with improving planning strategies student for the student's upcoming assessment tasks. Figure 8 prompts the students to start planning for an upcoming assessment while Figure 9 suggests the students reflect on their study strategies and seek to change their practises if they have not been successful in the past. This support is supplied in conjunction with other information such as that in Figure 3 that concurrently directs the students to the university's learning support services (Figure 4) who could assist with analysing and improving their study techniques.

The examples in this section indicated how the Checkpoints support the students' learning in several ways. The mathematical content, delivered through Topic pages, is interspersed with Checkpoints to break the mathematical content into smaller chunks. The Checkpoints also support the students to reflect on and re-focus their study practises by including a variety of alternative learning or study strategies.

5. Discussion

The discussion reports the feedback provided by the students and suggests takeaway points for teaching.

The students were asked whether they read the Checkpoints, and if the Checkpoints helped. Three replies were received:

(1) I did read the checkpoints, and found them to be a nice break from the Math content. They also helped me to reflect on my learning, and prompted me to write down areas I needed to work more on, and reassure[d] me if I was on track.
(2) I did read the checkpoints in the modules and found them helpful but maybe adding a few questions on the checkpoint could be helpful.
(3) I loved the checkpoints as a chance to take a breath and reflect on the contents

The feedback replies suggest that some students engaged with the Checkpoints' content, which is significant as the Checkpoints are not curriculum content and it might be reasonable for the students to avoid this content. The suggested improvement that the Checkpoints could include some mathematics questions highlights the balancing act between including more content and making the Checkpoints onerous tasks.

6. Takeaway points

The use of Checkpoints need not be limited to online subjects, nor to students with learning challenges. In the case discussed here, Checkpoints have supported the success of some students in an online mathematics subject, including those who were underprepared or affected by mathematics anxiety. However, integrating carefully considered extra-curricular content, such as Checkpoints, into any mathematics subject supports all students by providing regular breaks from learning and explicit opportunities for the students to reflect on their learning.

Checkpoints can contain content including signposts that support student engagement, motivational tips that support student wellbeing, learning tips that support student understanding, and assessment tips that support student performance. Some key questions to ask when preparing Checkpoints for online subjects include:

(1) What challenges affect the student's ability to engage with the subject content?
(2) How do I want the students to engage with the subject content?
(3) What can I do, considering the students' challenges, to support the student's engagement with the subject content?

7. Conclusion

Many students struggle with the mathematical demands of their university studies. These struggles have been exacerbated recently by the move to online learning resulting from the COVID-19 pandemic. It is important that academics remain aware of, and attend to, the additional needs of struggling students.

In light of the relative success of the transition to online learning, and the likelihood that the use of online learning would be extended, the continued development and use of Checkpoints would be worthwhile. This is particularly so for online subject delivery where students miss the interpersonal interactions to support learning that are normal in face-to-face classes. However, there is a need to ensure the Checkpoints are seen as supportive and not as extra work by balancing the benefit against the additional content.

In the case discussed here, the Checkpoints were intended to provide support for students learning mathematics online. The feedback suggested that the students engaged with the Checkpoints and that the Checkpoints did influence their learning, including prompting them to pause their study and reflect on their learning. These actions could have supported students impacted by mathematics anxiety and under-preparedness.

Disclosure statement

No potential conflict of interest was reported by the author.

References

Alamolhodaei, H. (2009). A working memory model applied to mathematical word problem solving. *Asia Pacific Education Review*, *10*(2), 183–192. https://doi.org/10.1007/s12564-009-9023-2

Ashcraft, M. H. (2002). Math anxiety: Personal, educational, and cognitive consequences. *Current Directions in Psychological Science*, *11*(5), 181–185. https://doi.org/10.1111/1467-8721.00196

Ashcraft, M. H., & Kirk, E. P. (2001). The relationships among working memory, math anxiety, and performance. *Journal of Experimental Psychology: General*, *130*(2), 224–237. https://doi.org/10.1037/0096-3445.130.2.224

Blazer, C. (2011). *Strategies for reducing math anxiety. Information capsule*. Volume 1102. Research Services, Miami-Dade County Public Schools.

Cornell, C. (1999). I hate math! I couldn't learn it, and I can't teach it!. *Childhood Education*, *75*(4), 225–230. https://doi.org/10.1080/00094056.1999.10522022

Dalitz, J. (2021a). *MA022: Essential maths: Module 1*. LearnJCU. https://learn.jcu.edu.au/

Dalitz, J. (2021b). *MA022: Essential maths: Module 2*. LearnJCU. https://learn.jcu.edu.au/

Dalitz, J. (2021c). *MA022: Essential maths: Module 3*. LearnJCU. https://learn.jcu.edu.au/

Exley, K., & Dennick, R. (2004). *Small group teaching: Tutorials, seminars and beyond*. Routledge.

Finlayson, M. (2014). Addressing math anxiety in the classroom. *Improving Schools*, *17*(1), 99–115. https://doi.org/10.1177/1365480214521457

Geist, E. (2010). The anti-anxiety curriculum: Combating math anxiety in the classroom. *Journal of Instructional Psychology*, *37*(1), 24–31. https://www.andrews.edu/sed/gpc/faculty-research/montagano-research/the-anti-anxiety-cur.pdf

Harper, N. W., & Daane, C. J. (1998). Causes and reduction of math anxiety in preservice elementary teachers. *Action in Teacher Education*, *19*(4), 29–38. https://doi.org/10.1080/01626620.1998.10462889

Jackson, C. D., & Leffingwell, R. J. (1999). The role of instructors in creating math anxiety in students from kindergarten through college. *The Mathematics Teacher*, *92*(7), 583–586. https://doi.org/10.5951/MT.92.7.0583

Khasawneh, E., Gosling, C., & Williams, B. (2021). What impact does maths anxiety have on university students? *BMC Psychology*, *9*(1), 1–9. https://doi.org/10.1186/s40359-021-00537-2

Kift, S. (2015). A decade of transition pedagogy: A quantum leap in conceptualising the first year experience. *HERDSA Review of Higher Education*, *2*(1), 51–86. https://altf.org/wp-content/uploads/2019/03/HERDSARHE2015v02p51-1-1.pdf

Lambie, I., & Law, B. (2020). *Teaching online during a pandemic: Pedagogical skills transfer from face to face support to online synchronous support provision* [Paper presentation]. 19th European Conference on e-Learning (ECEL), Reading.

MacGillivray, H., & Croft, T. (2011). Understanding evaluation of learning support in mathematics and statistics. *International Journal of Mathematical Education in Science and Technology*, *42*(2), 189–212. https://doi.org/10.1080/0020739X.2010.519801

Mann, G., & Willans, J. (2020). Monkey see, monkey do, that's not going to actually teach you. *Becoming a Self-Directed Learner in Enabling Mathematics Units. Student Success*, *11*(1), 55–66. https://doi.org/10.5204/ssj.v11i1.1462

Mann, V. E., Marshall, E. M., & Wilson, D. A. (2016). *Maths anxiety* [Paper presentation]. ADSHE 2016. https://maths.shu.ac.uk/mathshelp/Stats support resources/Maths anxiety/Publications/Mann,Marshall,Wilson(2016)MathsAnxietyADSHE.pdf

Marshall, E. M., Wilson, D. A., & Mann, V. E. (2018). Evaluating the effectiveness of maths anxiety awareness workshops. *Journal of Learning Development in Higher Education*, https://maths.shu.ac.uk/mathshelp/Stats support resources/Maths anxiety/Publications/Marshall, Wilson,Mann,(2017)Awareness_workshop_evaluation.pdf.

McGlaughlin, S. M., Knoop, A. J., & Holliday, G. A. (2005). Differentiating students with mathematics difficulty in college: Mathematics disabilities vs. no diagnosis. *Learning Disability Quarterly*, *28*(3), 223–232. https://doi.org/10.2307/1593660

Núñez-Peña, M. I., Bono, R., & Suárez-Pellicioni, M. (2015). Feedback on students' performance: A possible way of reducing the negative effect of math anxiety in higher education. *International Journal of Educational Research*, *70*, 80–87. https://doi.org/10.1016/j.ijer.2015.02.005

Perry, A. B. (2004). Decreasing math anxiety in college students. *College Student Journal*, *38*(2), 321–325. https://go.gale.com/ps/i.do?id=GALE%7CA119741942&sid=googleScholar&v=2.1&it=r&linkaccess=fulltext&issn=01463934&p=AONE&sw=w

Rylands, L. J., & Shearman, D. (2018). Mathematics learning support and engagement in first year engineering. *International Journal of Mathematical Education in Science and Technology*, *49*(8), 1133–1147. https://doi.org/10.1080/0020739X.2018.1447699

Sheffield, D., & Hunt, T. (2006). How does anxiety influence maths performance and what can we do about it? *MSOR Connections*, *6*(4), 19–23. https://doi.org/10.11120/msor.2006.06040019

Symonds, R., Lawson, D., & Robinson, C. (2008). Promoting student engagement with mathematics support. *Teaching Mathematics and Its Applications: International Journal of the IMA*, *27*(3), 140–149. https://doi.org/10.1093/teamat/hrn011

Taylor, J., & Galligan, L. (2006). Mathematics for maths anxious tertiary students: Integrating the cognitive and affective domains using interactive multimedia. *Literacy and Numeracy Studies*, *15*(1), 23–43. https://doi.org/10.5130/lns.v15i1.2025

Warwick, J. (2008). Mathematical self-efficacy and student engagement in the mathematics classroom. *MSOR Connections*, *8*(3), 31–37. https://doi.org/10.11120/msor.2008.08030031

Woodard, T. (2004). The effects of math anxiety on post-secondary developmental students as related to achievement, gender, and Age. *Inquiry*, *9*(1), 1–5. https://files.eric.ed.gov/fulltext/EJ876845.pdf

Young, C. B., Wu, S. S., & Menon, V. (2012). The neurodevelopmental basis of math anxiety. *Psychological Science*, *23*(5), 492–501. https://doi.org/10.1177/0956797611429134

Crafting connections in post-COVID classrooms: learning university mathematics through craft

Jeanette C. McLeod, Phillip L. Wilson, David Pomeroy and Julie Alderton

ABSTRACT

In 2020, after New Zealand emerged from its short-lived lockdown and students returned to university campuses, mathematicians Jeanette McLeod and Phil Wilson from the University of Canterbury taught a final-year mathematics course using craft. A part of most lectures included a hands-on craft activity to explore a key concept and to foster curiosity. This article describes the design and delivery of the course and reflects on the overwhelmingly positive student feedback. We suggest that the creation of, and interaction with, mathematical craft objects played a significant role in shaping these responses by generating feelings of engagement, reducing stress, and allowing tangible access to mathematical ideas.

1. Introduction

Maths Craft New Zealand is a non-profit public outreach initiative which brings maths to the masses by celebrating the links between mathematics and craft (www.mathscraftnz.org). Founded and run by authors Jeanette and Phil, Maths Craft's aim is threefold: to showcase the creativity and beauty inherent in mathematics through the medium of craft (Figure 1); to demonstrate just how much mathematics there is in craft; and to enable people to experience what it means to think like a mathematician, by doing craft. Since its inception in 2016, Maths Craft has run festivals and workshops across New Zealand, and has reached over 14,500 people, making it the largest mathematics outreach programme in the country.

While using craft to explain mathematics to the public, Jeanette and Phil realised its potential for teaching university students. In 2020, they took Maths Craft into their classroom and delivered half of a final-year university mathematics course, using craft. The course started in July 2020 after New Zealand emerged from its short-lived lockdown and students returned to university campuses for the first time since March 2020. In this article, we describe the design and delivery of the course, and the overwhelmingly positive student feedback collected in surveys and focus groups. Students reported that the creation of, and interaction with, mathematical craft objects played a significant role in generating feelings

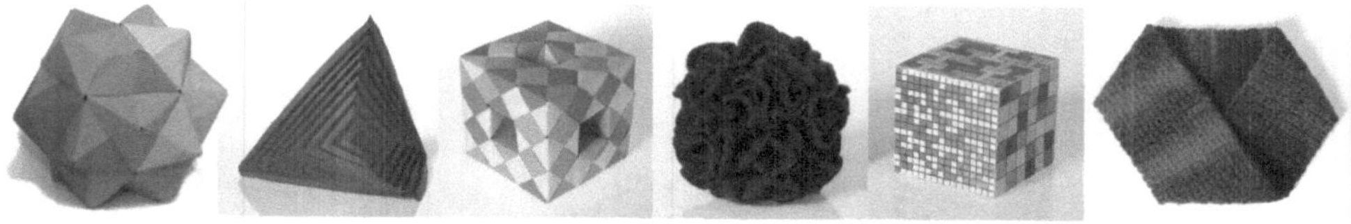

Figure 1. A selection of mathematical craft objects made by Maths Craft.

of engagement, reducing stress, and allowing tangible access to mathematical ideas. There is a growing body of research about the relationship between bodies, objects, and mathematics (de Freitas & Sinclair, 2014; Nemirovsky et al., 2020), but very little research on craft-based mathematics pedagogies in universities (Flórez & Mukherjee, 2020).

2. Design principles

The Maths Craft module was designed to be half of a 12-week final-year mathematics course which explores different perspectives of mathematics (the other half covered the history and philosophy of mathematics). This course is not compulsory for any degree, and with minimal course prerequisites, the mathematical backgrounds of the student cohort are diverse. Some students are majoring in mathematics and have taken many mathematics courses. Others have taken only a small number of lower-level courses in calculus and algebra. We had a cohort of 40 students in 2020.

Jeanette and Phil employed three key principles in the design of the Maths Craft module, drawing on their experiences of running Maths Craft events and teaching university mathematics:

(1) Capstoning each topic with a deep mathematical result;
(2) Weaving a story or narrative through each topic;
(3) Using craft as a scaffold for learning.

The module was split into four topics, each of which comprised three to five lectures, for a total of 16 50-minute lectures plus one Gallery Day (detailed later). For each topic, we first chose a major capstone theorem, for example, The Classification Theorem in the topology topic. We then identified the major milestones of each topic by considering both the mathematical story we wanted to tell, and the crafts we planned to use to illustrate them. The crafts became a scaffold for the topics, providing an unconventional but accessible entry point into each topic (such as starting topology with a hands-on exploration of the Möbius strip), and creating natural pauses for discussion and exploration. Woven throughout each topic were stories about the human and historical aspects of mathematics and the craft we were teaching. Finally, the module was designed to be accessible to students with a wide variety of mathematical backgrounds, from final-year mathematics majors to students from other disciplines who have not gone beyond second year mathematics.

Lectures consisted of a combination of hands-on craft activities, power-point slides, whiteboard work, class discussions, and the display and demonstration of various craft objects made by Maths Craft New Zealand. A significant aspect of the design of the course was the element of surprise and encounters with the unexpected. To this end, lecture slides

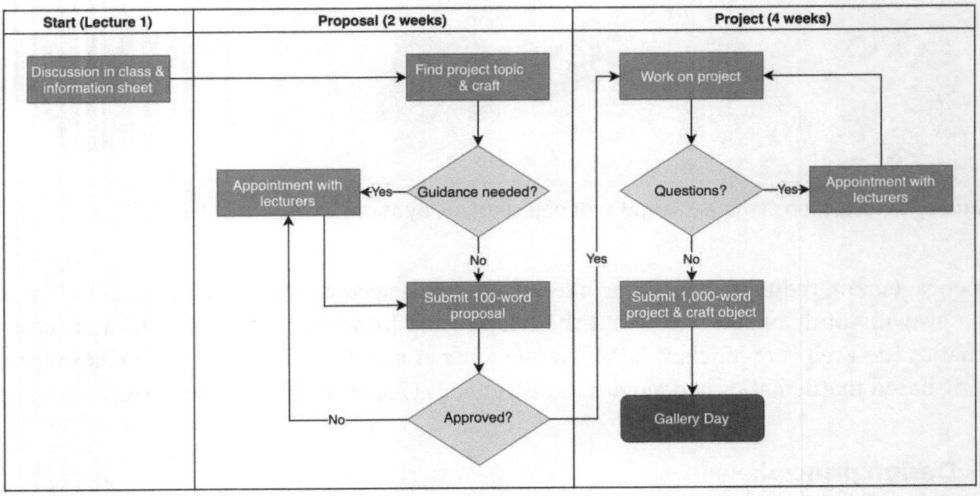

Figure 2. The Maths Craft project student workflow. Students were given detailed instructions on how to select a project and craft, and then given support and advice throughout the project.

were only posted online after lectures, instead of beforehand, as is usual at our university. The assessment of the Maths Craft module was worth 50% of the overall course grade and consisted of: a mark for participation (5% of the total grade); a take-home test (25% of the total grade); a craft project and associated written report about an aspect of mathematics illustrated by the craft project, or a mathematical analysis of the craft project itself (20% of the total grade) to be displayed at a (non-assessed) Gallery Day (see Figure 2). There were topic-based non-assessed homework questions.

3. Module contents

The Maths Craft module of the course was split into four topics, each of which involved both an in-depth study of the underlying mathematical theory, and the construction and exploration of craft objects in class. The structure of each topic is presented in Tables 1–4 to provide a framework for others wishing to create such a module.

3.1. Topology and the Möbius strip

Our first topic, Topology and the Möbius Strip, was designed to both provide a mathematical foundation for future topics, and to be enticing and challenging (see Table 1). To that end, our very first lecture began with students making and manipulating paper Möbius strips (Maths Craft New Zealand [MCNZ], n.d.-g). The topology of non-orientable surfaces is surprising and confounds intuition, while in contrast, the craft of constructing paper Möbius strips is highly accessible. This helped to illustrate that the crafts used in the course were not going to be hard and that the mathematics we can discover through craft is non-trivial. For example, although many students had previously seen Möbius strips, they had not explored them in any depth. So, when we asked them to predict the result of cutting a Möbius strip along its centre line before performing the activity, the result and

Table 1. Structure of the Topology and the Möbius strip topic. The capstone theorem is shown in bold.

Syllabus (5 lectures)	In-class Craft Activities
Introducing Topology and the Möbius strip. Non-orientable vs orientable surfaces	Making and manipulating a paper Möbius strip (MCNZ, n.d.-g)
Generalisation: strips with multiple twists	Creating and determining orientability of strips with multiple twists (odd vs even)
Generalisation: other non-orientable surfaces and their relationship to the Möbius strip	Demonstration of how the Möbius strip, torus, Klein bottle, and projective plane can be constructed from a rectangle of fabric, exploring chirality (MCNZ, n.d.-j), creating a projective plane from disposable napkins (Carroll, 1894)
Homeomorphism, surfaces, and boundary components	Determining the boundary components of the surfaces we have made: Möbius strip and projective plane
Genus and the Euler characteristic	Creating an origami cube and (small) triakis octahedron (MCNZ, n.d.-d, n.d.-e, n.d.-f) and determining their Euler characteristics
The Classification Theorem	Classifying the surfaces we made: Möbius strip, projective plane, and triakis octahedron

Figure 3. A sample of Jeanette's step-by-step photos taken while constructing Fortunatus's purse (projective plane).

their predictions were completely different, which surprised and delighted them. Having them discuss and explain the result, the disparity, and why their intuition had misled them, involved high-level mathematical thinking of a kind seldom modelled in other university mathematics courses.

In a later lecture on non-orientable surfaces, Jeanette and Phil playfully read a passage from *Sylvie and Bruno Concluded* (Carroll, 1894) in which a German mathematics professor describes the construction of Fortunatus's purse to a lady hemming handkerchiefs (complete with added social commentary on gender roles in the 1800s!) The students applauded at the end of the 'performance' and then made their own purses (which are essentially projective planes) from coloured disposable napkins and removable clips. To aid with this, Jeanette sewed her own plane and presented step-by-step instructions with photos (see Figure 3). The projective plane cannot be realised in 3-dimensional space without self-intersection, and is a difficult surface to visualise. However, creating the object helped students to understand not only this particular construction method (identifying the edge of a Möbius strip with a disc), but also how the limitations of 3-dimensional space manifest themselves.

This topic was presented as a quest to classify surfaces. What does it mean for two surfaces to be topologically equivalent (homeomorphic)? We established one property (such as orientability) only to discover that two surfaces sharing this property differed in other fundamental ways (especially evident when creating and handling the objects). The quest culminated in The Classification Theorem (see Table 1), one of the cornerstones of topology. Our approach meant that the entire journey was completed without using group theory (which students at the University of Canterbury seldom encounter). Instead, we gave students a mathematical framework for understanding the various topologically invariant properties of surfaces which allow their classification, as well why these invariants are both necessary and sufficient conditions.

3.2. *Map colouring*

The second topic, Map Colouring, took a more human look at mathematics. This time the quest for an answer to our focal problem (map colouring) was pursued through the lens of the various mathematicians who tried to solve the problem. We began with a look at the long and controversial history of The Four Colour Theorem (Wilson, 2014). Then, combining this with topology from our first topic, we considered maps on other surfaces (classified by genus), culminating with the Map Colour Theorem. Students' intuition was again challenged when it was revealed that the Map Colour Theorem, although applicable to many other seemingly more complex surfaces, was much easier to prove than the Four Colour Theorem (Table 2). And while some students had heard of the Four Colour Theorem, none were aware of the Map Colour Theorem or its equally intriguing history. Experiments with the Möbius strip and trying to ascertain how many colours are required for a map colouring on this surface, were a source of much lively discussion amongst the students, and some creative (and hilarious) maps were drawn and bold conjectures made.

Table 2. Structure of the Map Colouring topic. The capstone theorem is shown in bold.

Syllabus (3 lectures)	In-class Craft Activities
Defining the Four Colour Problem and explaining why it is hard	Drawing: attempting to solve the Five Princes Problem
The history of the Four Colour Theorem: the search for a proof, its discovery and ramifications	Colouring a famous (fake) 'counterexample' (MCNZ, n.d.-h)
Counting colourings	Using a real-life craft object (crocheted blanket) as an example
The Map Theorem: maps on other surfaces	Creating and colouring minimal 4-, 5- and 6-colourable maps on Möbius strips

3.3. *Fractals*

The third topic was fractal geometry. In the first lecture of this topic, students collaborated on building an approximate fractal sculpture made from business cards (MCNZ, n.d.-a, n.d.-i). The craft skills required for building the Menger Sponge are minimal, involving no more than folding two parallel lines in business cards followed by a simple iterative assembly method (see Figure 4). While most students began working on a Menger Sponge alone, they all spontaneously formed groups in which either individuals simultaneously made cubes, or divided the labour into an 'assembly line' with each student working on a different aspect of the construction. Finally, the groups merged, unprompted, into a whole-class effort to complete the sculpture.

Table 3. Structure of the Fractals topic. The capstone theorem is shown in bold.

Syllabus (4 lectures)	In-class Craft Activities
History and key concepts	Building a Menger sponge from business cards
Scaling, Hausdorff dimension, and famous fractals	Drawing Cantor dust, Koch curve, and Sierpinski Triangle. Knitted Sierpinski Triangle (shown in class)
Minkowski dimension	Box counting on maps, kirigami triangles
Chaos, dynamical systems, strange attractors, the Lorenz manifold, and **Smale's 14th Problem**	Using images of a real-life craft object: Hinke Osinga's crocheted Lorenz Manifold (Osinga & Krauskopf, 2004)

The iterative construction method of the Menger Sponge served as a scaffold for key concepts in subsequent lectures, all of which involved the definition and analysis of iteratively defined fractal objects (Table 3). Many of these objects are visually appealing and can easily be drawn by hand; students drew Cantor Dust, the Koch Curve, and the Sierpinski Triangle (they also examined a beautiful Sierpinski Triangle knitted by Jeanette). The study of each fractal contributed to a developing story about self-similarity and fractal dimensions. The elements of this story were embedded in an historical context which was often far removed from the modern discovery of fractals. For example, when Cantor discovered Cantor Dust he was exploring the foundations of set theory. The human side of mathematics was once again highlighted both by the quests of the protagonists and their often difficult but inspiring lives.

The culmination of this topic was an exploration of Smale's 14th Problem. Smale's Problems (Smale, 1998) are a list of 18 important mathematical problems unsolved at the time of their publication. The 14th problem asks whether or not a chaotic system known as the Lorenz system describes a mathematical object with fractal properties. This mathematical object is a manifold, which recalled an understanding of manifolds from the topic Topology and the Möbius Strip. The complicated shape of the Lorenz manifold was computed in the late 1990s by Hinke Osinga and Bernd Krauskopf, who transformed their numerical method into a crochet pattern and made the first crocheted Lorenz manifold (Osinga & Krauskopf, 2004). While we did not have (and could not make!) this object in class, Phil

Figure 4. Left: Jeanette and Phil's demonstration Level 1 Menger sponge. Right: The students' collaborative Level 1 Menger Sponge under construction during the lecture.

used images of it to describe Smale's 14th problem and the important connection between fractals and chaos.

3.4. Knot theory

The final topic, knot theory, began with experiments using lengths of coloured string. Students tied knots before Phil led the class through physical explorations of them. When is one knot simpler than another? Can one knot be manipulated to look like any other? Can it be manipulated to look like its mirror image? These experiments quickly and concretely illustrated some of the main ideas of knot theory, which were explored over subsequent lectures (see Table 4). Experiments with string were used throughout, and Maths Craft handouts on French knitting mathematical knots (MCNZ, n.d.-b, n.d.-c) were made available. Because knot theory is a branch of topology, many links were made between this topic and the topic Topology and the Möbius Strip. In particular, the capstone theorem for Knot Theory was a classification theorem. Although knots are familiar everyday objects, students were delighted that many simple statements about them remain as conjectures.

Table 4. Structure of the Knot Theory topic. The capstone theorem is shown in bold.

Syllabus (4 lectures)	In-class Craft Activities
Introducing knot theory and basic concepts	Tying and manipulating string knots
Alternating, composite, prime, and amphichiral knots	
Ambient isotopies and the Reidemeister moves	Learning Reidemesiter moves via manipulating string, finding equivalent projections
Knot invariants – crossing number, tricolourability	Counting crossings of string knots
Thurston's Knot Classification Theorem	Knitted torus knots, crocheted hyperbolic plane (shown in class)

4. Methods

We devised a small-scale study to evaluate students' experiences of the course and their levels of cognitive, emotional, and behavioural engagement (Pilotti et al., 2017). The methods included student questionnaires ($n = 18$), one focus group interview with three students, and notes from observations of four lectures and the Gallery Day. The evaluation was approved by the University of Canterbury Educational Research Human Ethics Committee.

5. Discussion and implications

We argue that making and interacting with craft objects contributed to unusually high levels of engagement in mathematics learning and reduced learning-related stress. We suggest that other university mathematics educators consider whether making and/or interacting with mathematical objects, could yield similar benefits in their courses, especially post-COVID. This short discussion is our initial reflection on the ways in which the design principles (mathematical depth, narrative, and craft) affected students' engagement in learning and *how* they learned (Figure 5).

Figure 5. The Gallery Day. Top: Students explain their creations to each other. Bottom left: The Huffman Tower. Bottom right: a catenoid.

Students reported and demonstrated high levels of overall engagement with this course. Out of the class of 40, 18 students completed the questionnaire, and all enjoyed the course more (7 students) or a lot more (11 students) than other university mathematics courses. Jeanette and Phil noticed markedly higher lecture attendance than is usual for a final-year course, which ran counter to a trend of low attendance post-lockdown. Almost all students attended the Gallery Day even though it was optional and ungraded, and all who attended gave an impromptu talk about their objects (see Figure 5). Most students invested more effort in making their object than Jeanette and Phil expected, even though the quality of the object was not graded. One student who built an enormous 3D Sierpiński triangle explained that for her it was a matter of 'go big or go home.'

Making crafts was the most frequent reason students gave in the questionnaire for their enjoyment of the course. Students often associated their enjoyment of crafting with a qualitative shift in how they were able to approach deep mathematical ideas. In the focus group, a student described the way in which physically manipulating a Möbius strip supported her learning of topology concepts:

> It was really cool to be like 'here is the basic gist of it', play around with some crafts and then be like 'here is what it means formally'. It gives you a good base for understanding it, because even if you don't understand the formal definition, you still have a good grasp on what it actually means. Whereas maybe in a formal class, you would start with the formal definition and just go from there and be like 'I have no idea what this means'.

Students attributed learning benefits of crafting not only to the conceptual scaffolding provided by physical representations, but also to the 'think time' and stress reduction of completing a repetitive and absorbing task. Discussing physical objects, one student explained:

> I feel like [craft] makes you more excited about it [mathematics] and that you want to learn about it. I think I learned a lot just because I wasn't stressed about absorbing heaps of maths. I was just happy to just relax and sit and learn more.

Such perceived stress reduction seems especially pertinent emerging from the fear, isolation, and uncertainty of lockdown.

6. Conclusion

We designed a 6-week module for a final-year university mathematics course using principles derived from the mathematics outreach programme Maths Craft New Zealand. Each topic led to a deep mathematical result, told a story, and used craft objects to support learning. Students were highly engaged in the course and reported that the approach made mathematics more accessible and intuitive, and mathematics learning less stressful. Such outcomes are all the more pertinent when emerging from the uncertainty and isolation of COVID-related lockdowns.

Disclosure statement

No potential conflict of interest was reported by the authors.

ORCID

Jeanette C. McLeod http://orcid.org/0000-0003-1729-5100
Phillip L. Wilson http://orcid.org/0000-0002-4563-9399
David Pomeroy http://orcid.org/0000-0003-3187-4942
Julie Alderton http://orcid.org/0000-0002-1408-7640

References

Carroll, L. (1894). *Sylvie and Bruno concluded*. Macmillan. http://www.gutenberg.org/files/48795/48795-h/48795-h.htm, https://www.gutenberg.org/files/48795/.

de Freitas, E., & Sinclair, N. (2014). *Mathematics and the body: Material entanglements in the classroom*. Cambridge University Press.

Flórez, R., & Mukherjee, A. (2020). Classroom experiences in introducing students to conjectures, exploration and visual proofs using experiments on the Möbius strip. *PRIMUS*, *30*(1), 50–66. https://doi.org/10.1080/10511970.2018.1506530

Maths Craft New Zealand. (n.d.-a). *Building the Menger Sponge.* http://www.mathscraftnz.org/s/Handout-Menger.pdf.

Maths Craft New Zealand. (n.d.-b). *French knitting basics.* http://www.mathscraftnz.org/s/Handout-French_Knitting.pdf.

Maths Craft New Zealand. (n.d.-c). *French knitting mathematical knots.* http://www.mathscraftnz.org/s/Handout-French_Knitting_Knots.pdf.

Maths Craft New Zealand. (n.d.-d). *Making an origami Sonobe cube.* http://www.mathscraftnz.org/s/Handout-Sonobe_Cube.pdf.

Maths Craft New Zealand. (n.d.-e). *Making an origami Sonobe octahedron.* http://www.mathscraftnz.org/s/Handout-Sonobe_Octahedron.pdf.

Maths Craft New Zealand. (n.d.-f). *Making an origami Sonobe unit.* http://www.mathscraftnz.org/s/Handout-Sonobe_Unit.pdf.

Maths Craft New Zealand. (n.d.-g). *Making and manipulating a Möbius strip.* http://www.mathscraftnz.org/s/Handout-Mobius_Strip.pdf.

Maths Craft New Zealand. (n.d.-h). *Map colouring and the Four Colour Theorem.* http://www.mathscraftnz.org/s/Handout-Colouring-Four_Colour_Theorem.pdf.

Maths Craft New Zealand. (n.d.-i). *Menger Sponge cladding.* http://www.mathscraftnz.org/s/Menger_Cladding_A4-9up.pdf.

Maths Craft New Zealand. (n.d.-j). *The heart of a Möbius.* http://www.mathscraftnz.org/s/Handout-Mobius_Strip-Hearts.pdf.

Nemirovsky, R., Ferrara, F., Ferrari, G., & Adamuz-Povedano, N. (2020). Body motion, early algebra, and the colours of abstraction. *Educational Studies in Mathematics, 104*(2), 261–283. https://doi.org/10.1007/s10649-020-09955-2

Osinga, H. M., & Krauskopf, B. (2004). Crocheting the Lorenz manifold. *Mathematical Intelligencer, 26*(4), 25–37. https://doi.org/10.1007/BF02985416

Pilotti, M., Anderson, S., Hardy, P., Murphy, P., & Vincent, P. (2017). Factors related to cognitive, emotional, and behavioral engagement in the online asynchronous classroom. *International Journal of Teaching and Learning in Higher Education, 29*(1), 145–153. https://www.isetl.org/ijtlhe/pdf/IJTLHE2503.pdf.

Smale, S. (1998). Mathematical problems for the next century. *Mathematical Intelligencer, 20*(2), 7–15. https://doi.org/10.1007/BF03025291

Wilson, R. (2014). *Four colours suffice*. Princeton University Press.

The pandemic as a catalyst for rethinking active learning practices in technology intensive instructional environments

David Christopher Webb

ABSTRACT

Even though K-12 mathematics educators have embraced the role of graphing calculators and computers, student use of technology in tertiary mathematics has been slow to implement in the USA context. Enter March 2020, a global health crisis, and the shift to online and remote learning. The immediate and dramatic cessation of in-person modes of instruction was experienced at all levels of education and in all disciplines. With respect to tertiary mathematics, student use of instructional technology shifted from optional use of homework websites, digital discussion boards and flipped classrooms to an absolute necessity to support student participation in mathematics courses. Our research group has been studying the process of departmental change towards active learning in the undergraduate calculus sequence. During the COVID-19 pandemic, we worked with twenty-three university mathematics departments in the United States as part of a Networked Improvement Community (NIC). During the pandemic, we observed the adaptation of instructional practices to sustain student interaction in remote virtual environments. This paper highlights changes observed within our NIC regarding attempts to sustain active learning during the global pandemic. In addition, a NIC can provide the necessary community for grappling with challenging problems at any time.

1. Introduction

Undergraduate mathematics education and instructional technology have had a complicated history in the United States. While K-12 mathematics educators in the US were embracing the role of graphing calculators and computers during the early 1990s (NCTM, 1989), most tertiary mathematics instructors and mathematics departments restricted student use of technology in introductory mathematics courses. Enter March 2020, a global health crisis, and the rapid shift to various iterations of online and hybrid teaching. The immediate and dramatic cessation of in-person modes of instruction was experienced at all levels of education and across all disciplines.

Prior to the COVID-19 pandemic, emerging examples of instructional technology use in tertiary mathematics were observed within our own mathematics department and elsewhere. In this article, instructional technology is defined as the regular use of computer hardware and software to support teaching and learning (and not incidental use of webpages or calculators). University administrators and select instructors have been advocating for increased use of online courses to provide students flexible options for course and degree completion. With a larger instructional reach, one university could serve students regionally, nationally, and internationally.

Even among mathematics departments that did not offer online courses or virtual instruction, technology was used over the past decade to support to student learning. Two decades ago in the United States, a number of mathematics departments experimented with an approach where students worked individually to complete instructional modules and assessments on desktop computers at their own pace in a proctored computer lab (Twigg, 2011). On the other hand, there has been widespread use of homework websites by mathematics departments: students log into a personal account, complete problem sets, and receive individualized support through low-grade 'feedback' such as whether a response was correct or incorrect. Still, prior to the pandemic, tertiary mathematics instructors did not use technology as a resource to support day-to-day instruction, communication, or student inquiry.

In March 2020, the immediate shift to online and remote forms of tertiary mathematics instruction throughout much of the world was dramatic and overwhelming to instructors and students. During this, and through the subsequent academic year, we were engaged in a research project involving twenty-three tertiary mathematics departments who were investigating ways to implement and sustain active learning practices (Freeman et al., 2014; Laursen & Rasmussen, 2019). This article focuses on how the COVID pandemic impacted the progress of mathematics departments in implementing active learning and on the classroom practices of instructors of high enrollment introductory mathematics courses – i.e. the Precalculus through Calculus 2 (P2C2) sequence.

2. Context

Since 2016, our research group has been studying the process of departmental change towards active learning in the undergraduate calculus sequence as part of the Student Engagement in Mathematics through an Institutional Network for Active Learning (SEMINAL) project. The mathematics departments we have worked with to date were selected based on their interest in initiating or success in sustaining departmental change initiatives designed to improve calculus programs using active learning. The SEMINAL project's primary research question is: What conditions, strategies, interventions, and actions at the departmental and classroom levels contribute to the initiation, implementation, and institutional sustainability of active learning in the undergraduate calculus sequence across varied institutions?

The latter phases of the SEMINAL project occurred during the COVID-19 pandemic. Nearly all members of the research team and participants involved with SEMINAL were teaching courses when universities across the US transitioned to online or remote forms of instruction in late March 2020. The intensified workload, anxiety and sense of uncertainty was experienced by everyone involved with SEMINAL. The physical safety of students, self, and colleagues was a concern, and this was followed soon after by concerns over mental well-being. As tumult shifted to acceptance of the situation, we soon recognized that we had an unanticipated opportunity to support each other in problem-solving the instructional design challenges through the Networked Improvement Community (NIC; Bryk et al., 2011; Martin & Gobstein, 2015) we developed in the SEMINAL project.

Among the data sources used to address the primary research question for SEMINAL were site visits in which we conducted interviews with university administrators, mathematics department chairs, and key faculty including instructors of P2C2 courses. After March 2020, most universities were deserted, and travel was not permitted. So, for thirteen of the partner institutions, we conducted virtual site visits and used video conferencing technology to record interviews of key personnel.

3. An emergent Networked Improvement Community

Much has been written about the purpose, goals, and functions of a Networked Improvement Community (NIC; Bryk et al., 2011; Gomez et al., 2016; Martin & Gobstein, 2015; Russell et al., 2017). The SEMINAL NIC was organized to address a common goal among the initial twelve, and later twenty-three, partner institutions. SEMINAL NIC members were interested in ways to initiate, scale, and sustain active learning practices in P2C2 courses in their respective mathematics departments. In addition to using elements of improvement science to iteratively adapt and refine norms, structures, resources and practices, the network supports the regular sharing of ideas (e.g. methods for coordinating common assessments) and resources (e.g. group worthy tasks, conceptual assessments). Through regular dissemination of local experimentation, NIC partners begin see how similarities among mathematics departments could be generalized as potential design principles.

4. Pre-pandemic progress with active learning

Prior to the pandemic the mathematics departments participating in SEMINAL were all making progress toward developing and disseminating curricular resources that could be used by all P2C2 instructors to support active learning. Faculty designated as course coordinators conducted meetings with other instructors to develop a shared understanding of active learning in mathematics (ALM) that included students' 'deep engagement with meaningful mathematics and collaborative processing of mathematical ideas' (Laursen & Rasmussen, 2019, p. 138). Mathematics departments also used different strategies to implement ALM to account for local affordances and constraints. Improving course coordination structures and communication was used by at least half of the mathematics departments to spread and sustain ALM implementation. One-third of the mathematics departments used undergraduate learning assistants to catalyze and support group discussions. Two mathematics departments focused on the integration of technology: one used an instructional approach where instructors created videos of lectures for students to watch prior to class so that in-person lectures were replaced by group work and student discussions about select mathematics tasks and concepts. This approach has been referred to elsewhere as the flipped classroom model (Bergmann & Sams, 2012). Instructors in the other department, to create more active engagement in large enrollment lectures, used interactive notes and polling apps to pose questions to students in the middle of a lecture. Students responded using their digital devices which allowed the instructor to formatively assess the range of student responses to a task.

Following an in-person SEMINAL workshop in June 2018, we initiated regular virtual monthly meetings during each academic year which were later branded as 'NICcasts'. A second in-person workshop was conducted in June 2019 in which SEMINAL partners and faculty from other research projects shared their progress with implementing active learning. Monthly NICcasts continued through the subsequent academic year. Therefore, prior to the onset of the pandemic, a sense of community and mutual interest in each other's work was developed through the SEMINAL NIC. The community that was developed through a combination of in-person and virtual meetings to later support each other through the pandemic was serendipitous and its value cannot be understated. Overall, during a period of two years and leading up to the COVID pandemic, almost all mathematics departments participating in SEMINAL had made significant progress toward enacting their proposed strategies and accomplishing their stated goals for implementing ALM.

5. Rethinking ALM in virtual and online environments

The move to online and remote teaching during the middle of March 2020 motivated us to organize ad hoc NICcasts to articulate the challenging circumstances we and our students were facing and name the problems we were encountering so that we could problem solve the most urgent concerns and propose strategies for re-examining ways to incorporate ALM practices in remote instructional settings.

The first of these NICcasts, hosted on March 20, 2020, presented participants with the following discussion prompt (Figure 1).

In response to this prompt, breakout groups noted the following challenges:

- Takes so much more instructor time to get online set up and to do the logistics of moving into breakout rooms, etc.

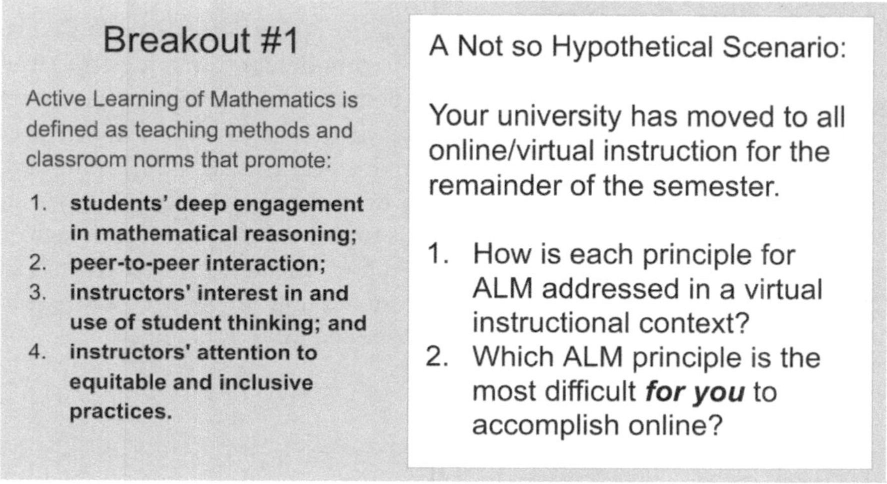

Figure 1. NICcast prompt for discussions groups (March 20, 2020).

- How to handle the whole class discussion, not just the breakout rooms. The concern is that the whole class discussion will not be as free-flowing as in-person
- Can feel very odd to drop into breakout groups and just observe and not talk, despite this being how it worked to observe groups in-person.
- What about access issues? Do students have devices to get online ... ?

On a more positive note, breakout groups also noted how virtual platforms can 'provide opportunities for students to work on their ability to communicate math' and saw some reticent students who were starting to engage more frequently in classroom discussions. Instructors who were required to move to asynchronous, online instruction planned to have groups of students record presentations on assigned topics to upload and share with the rest of the class for peer review. Despite the intensification of work and increased anxiety among instructors during the first month of the pandemic shutdowns, creative and thoughtful approaches were proposed to sustain active learning in their P2C2 courses. Having a forum to share resources, practices and proposals helped ease the burden of instructors trying to resolve the challenges on their own.

Additional NICcasts hosted during the remainder of the semester, before and through the 2020–2021 academic year allowed participants to showcase ways to teach mathematics using a makeshift whiteboard at home; how to use video conferencing features like chat, polls and breakout rooms to support student communication, engagement, and collaboration on mathematics tasks. Despite the turmoil, we recognized that features of video conferencing could be beneficial to instructors and students – e.g. seeing the name of classmates, microphones and volume controls allowed instructors and students to hear more easily what other students were saying, classes could be recorded and automatically transcribed for students who could not attend a session, and so on. Other NICcast topics included: the design and use of assessments, course coordination, designing equitable and inclusive ALM classrooms, students' experiences with ALM, and characteristics of supportive classroom environments for students. The NIC also collaborated in the development of a Remote Teaching & Learning FAQ that summarized key ideas presented in the NICcasts (SEMINAL project, 2020).

Unfortunately, many instructors were not able to attend SEMINAL NICcasts and remained disconnected from locally organized opportunities for sustaining active learning during the pandemic or the opportunity to learn about resources developed by other mathematics departments. As we found out later during virtual site visit interviews, instructors from five mathematics departments described a move away from active learning and how challenging it was to engage students in online or hybrid environments: '[ALM] is almost non-existent to be honest ... you can't let them talk to each other or be close to each other'. An instructor in another department summarized their situation as, 'COVID has stalled us. COVID has set us backward ... I've gotta figure out how to create an active learning environment online, and I don't know where to begin on that'.

6. Future implications

Given the educational challenges teachers and students experienced during the global pandemic, it is easy to dismiss the event as an anomalous crisis that we can soon forget as we return to pre-pandemic practices. However, the upheaval of the norms and traditions in tertiary mathematics motivated administrators and instructors to rethink how ALM could be sustained in technology intensive instructional contexts. The process of rethinking ALM in different instructional contexts also informs what tertiary mathematics classrooms and instruction might be going forward. Redesigning courses, resources and practices to support ALM compelled instructors to re-examine principles for ALM and analyse the affordances and constraints of different instructional contexts. As one university administrator surmised during a recent virtual site visit:

> [the pandemic] has opened the door for discussion, which I think is very serious here... I have no understanding, frankly, of why we will ever have 300 students sitting in an auditorium lecture again. To me, that makes absolutely no sense given the technical achievements of our faculty in general... The idea that we would do that it's just wasteful with wasted space, wasted facilities, wasteful of student time. Frankly, it is inflexible in terms of faculty time. It's just not a model that can be sustained any longer and justified any longer.

7. Takeaway

Even though the SEMINAL NIC was formed prior to the pandemic to promote the implementation and expansion of active learning, during the pandemic the SEMINAL NIC provided a place to commiserate and share creative solutions to rethink tertiary mathematics instruction. The formation of a NIC can be an invaluable way to address a common problem as participation in regular meetings provides opportunities to discuss and make sense of issues related to the problem of interest. A NIC also can be trusted place to share resources and support colleagues in the midst of unexpected challenges.

Acknowledgements

All research conducted in this study was approved by the University of Colorado Boulder Institutional Review Board (IRB Protocol #18-0099). As described in the approved research protocol, all subjects interviewed for this research read and signed informed consent documents prior to the administration and recording of the interviews.

Disclosure statement

No potential conflict of interest was reported by the author.

Funding

This material is based upon work supported by the National Science Foundation, USA [grant numbers DUE-1624610, 1624628, 1624639, 1624643]. Any opinions, findings, and conclusions or recommendations expressed in this material are those of the author(s) and do not necessarily reflect the views of the National Science Foundation.

ORCID

David Christopher Webb http://orcid.org/0000-0001-6661-4679

References

Bergmann, J., & Sams, A. (2012). *Flip your classroom: Reach every student in every class every day*. International Society for Technology in Education.

Bryk, A. S., Gomez, L. M., & Grunow, A. (2011). Getting ideas into action: Building networked improvement communities in education. In M. T. Hallinan (Ed.), *Frontiers in sociology of education* (pp. 127–162). Springer.

Freeman, S., Eddy, S. L., McDonough, M., Smith, M. K., Okoroafor, N., Jordt, H., & Wenderoth, M. P. (2014). Active learning increases student performance in science, engineering, and mathematics. *Proceedings of the National Academy of Sciences, 111*(23), 8410–8415. https://doi.org/10.1073/pnas.1319030111

Gomez, L. M., Russell, J. L., Bryk, A. S., LeMahieu, P. G., & Mejia, E. M. (2016). The right network for the right problem. *Phi Delta Kappan, 98*(3), 8–15. https://doi.org/10.1177/0031721716677256

Laursen, S. L., & Rasmussen, C. (2019). I on the prize: Inquiry approaches in undergraduate mathematics. *International Journal of Research in Undergraduate Mathematics Education, 5*(1), 129–146. https://doi.org/10.1007/s40753-019-00085-6

Martin, W. G., & Gobstein, H. (2015). Generating a networked improvement community to improve secondary mathematics teacher preparation: Network leadership, organization, and operation. *Journal of Teacher Education, 66*(5), 482–493. https://doi.org/10.1177/0022487115602312

NCTM. (1989). *Curriculum and evaluation standards for school mathematics.*

Russell, J. L., Bryk, A. S., Dolle, J., Gomez, L. M., LeMahieu, P., & Grunow, A. (2017). A framework for the initiation of networked improvement communities. *Teachers College Record, 119*(7), 1–36. https://doi.org/10.1177/016146811711900501

SEMINAL Project. (2020, December 18). Remote teaching & learning FAQ. https://tinyurl.com/SEMINALCOVIDFAQ.

Twigg, C. A. (2011). The math emporium: Higher education's silver bullet. *Change: The Magazine of Higher Learning, 43*(3), 25–34. https://doi.org/10.1080/00091383.2011.569241

The unexpected consequences of replacing corridor chats with an online community of practice

Anita L. Campbell, Tracy S. Craig, Batseba Mofolo-Mbokane and Pragashni Padayachee

ABSTRACT

We are four academics teaching mathematics at three different higher education institutions over two continents. Through fortnightly meetings to research how our practices changed as a result of the Covid-19 lockdowns, we formed a community of practice. The process of articulating our experiences and ideas, and reflecting on the recorded meetings and the written homework we set ourselves, was reassuring and motivational. We identified five assessment propositions to guide our mathematics teaching, which we summarize in this paper. A major proposition of assessment we embraced was the use of open book (or open-internet) assessment as a way to test for mathematical understanding. Our community of practice interactions influenced our teaching and assessment practices. We learnt more deeply about assessment by interrogating each other's work, observing and identifying misconceptions or errors (made by ourselves and others), and learning different ways of solving problems through discussion. We noted that sustaining the community of practice required comfort in being confronted and criticized. The unexpected consequence of our community of practice was the push it gave us to think about why we teach what we teach, assess how we assess, and how we can make both more relevant to a changing world.

1. Coming up with our own corridors

As academics, we understand that learning is seamless and occurs effectively in collaboration with others through brainstorming ideas, offering different perspectives, challenging, and critiquing. We advocate the value of collaboration to our students but often undermine its value for our own learning and practice, conveniently slipping into working in silos amidst the myriad other academic activities assigned to us. Our isolation due to the Covid-19 pandemic provided the incentive to create online opportunities to replace corridor chats with colleagues about our experiences, challenges, and ideas. Through WhatsApp chats, we created a virtual, dual-hemisphere corridor that sparked regular meetings on Zoom where we could find reassurance and new ideas for our developing online and blended teaching practices.

We are all female academics who have worked in higher education for more than 20 years and we have all taught engineering mathematics. We regularly attend the Southern Hemisphere Delta Conferences on the Teaching and Learning of Undergraduate Mathematics and Statistics and were all part of the organizing committee for the Elephant Delta conference in 2015. A measure of our commitment was that we remained members of the 'Elephant Delta committee' WhatsApp group. This WhatsApp group was our initial mode of communication, soon replaced by a new group 'Collaboration Engineering Maths' in February 2021. The prospect of a collaboration on a future conference provided further incentive for fortnightly meetings on Zoom.

When we first met, we were all based in South Africa, at three different universities; now we are spread across two universities in South Africa and one in the Netherlands. At the start of 2020, three of us were teaching calculus to first- and second-year engineering students, while one taught calculus content to third-year Bachelor of Education students. Class sizes ranged from 24 to 150 and we each taught one or two classes per semester. The mode of teaching and learning was mainly face-to-face lectures daily (for three of us in South Africa) or weekly (in the Netherlands) followed by two to four hours of problem solving, collaborative tutorials once a week. From mid-March until the end of 2020, our classes were fully online, and in 2021 two of us in South Africa had face-to-face tutorials. In 2020, a low-technology policy was advocated in South Africa to minimize disparities due to device and internet access limitation, therefore we avoided or minimized synchronous video meetings.

Our common values, experiences, and concerns regarding our teaching gave us assurance that we could speak openly and honestly with each other. Our discussions focussed less on our teaching content, but on assessment methods that worked for us individually during the Covid-19 pandemic. Reflecting on these discussions influenced the way we continue to teach and assess, in the same way that chats with colleagues in physical corridors benefitted us in the past. Gradually we recognized that we were working as a community of practice (CoP).

In the next section, we define a CoP and discuss the logistics of our CoP. Thereafter we discuss the value we found from listening to and questioning each other's practices and applying the shared ideas in our classes. We summarize five assessment propositions that emerged. Finally, we suggest practices that can help to sustain a CoP.

2. Defining a community of practice (CoP)

Throughout the ages people have joined together to share ideas and resources, solve problems, and invent new territories. Communities of practice are formed by groups of individuals who share a commonality and regularly interact to work together on that issue or concern (Wenger, 2000). There are three required components of CoPs (Wenger-Trayner & Wenger-Trayner, 2015):

(1). There needs to be a *domain*. Our domain was that we were university calculus teachers.

(2). There needs to be a *community*. Our fortnightly meetings on Zoom and collaborative writing on shared Google documents provided the community.
(3). There needs to be a *practice*: Over time we developed a shared repertoire of experiences and resources for online assessment which informed our teaching practice.

In a CoP, members participate voluntarily and share knowledge. Each participant has their own competencies or areas of strength that become more valuable when shared in the community. The unusual circumstances created by the pandemic challenged our mathematics teaching and learning practices and required that we innovate, develop, and create new ways of doing. Our CoP interactions highlighted the need to move away from our traditional ways of teaching and to consider appropriate new methods for a not-before-experienced situation. Using online platforms (WhatsApp and Zoom) made it easy to share and store resources to return to for later reflection. At its core, our CoP allowed us to redefine knowledge, and not simply transfer knowledge from one to another.

Lave and Wenger (1991) stress the multiple values in social learning related to a CoP. Reaching out to each other provided immediate value because we could voice common concerns, take comfort knowing others were experiencing the same, and support each other. In our many discussions, brainstorming and critiques of our practices, we came up with alternatives to teach and assess and this provided the potential value. Between our meetings, we formulated these ideas for our contexts, and these became the applied value. Actualizing these ideas into our classrooms, constituted the realized value. In addition, there is value to us and our wider communities – departments, faculties, and universities we are part of – in sharing our CoP experiences and knowledge.

We note some factors that helped our CoP to function:

- Commitment to each other to show up;
- Feeling comfortable to be authentic and to voice our opinions, feelings, and criticisms;
- Reflexively to embrace the views of the others of the others on the CoP;
- No fear of judgement;
- Homework to complete between meetings.

Our homework was reading and writing about our experiences with assessment in our courses. Once we decided to work towards research we could present at the 2021 Herenga Delta conference, our homework was more focused writing, or gathering and analysing questions.

3. The value of our CoP

Our CoP meetings met all of the five ways that corridor chats can improve teaching as identified by Thomson and Trigwell (2018): managing practicalities; sharing tips to engage students; reassurance on teaching plans; venting teaching-related frustrations; and sparking transformation of teaching practices. There were differences in our cohorts (for example, class sizes, students' access to technology and bandwidth, requirements to standardize courses with colleagues) and our ways of working, yet value could be found for our own contexts.

3.1. Assessment propositions to guide our teaching

During our fortnightly online 'corridor chats' we, unsurprisingly under the circumstances, found a common interest in assessment. We were interested to know what values and beliefs underpinned our collective approach to assessment. We summarize our perceptions of the benefits that outweigh negative effects, and give practical examples, in the five assessment propositions below. They are: open book assessment has value; verbalizing helps learning; assessment can promote self-directed learning; assessment can develop higher order thinking skills (HOTS); and assessment as learning and for mastery learning benefits students.

The sudden shift to an online educational environment forced us to adopt **open book assessment**, a shift we have come to recognize as valuable. Despite a long history of research on open book assessment (for example, Cnop & Grandsard, 1994; Croasdale, 1973), open book assessment is not widely used in mathematics (Iannone & Simpson, 2021) and hence is controversial. Yet, open book assessment closely resembles the real world of work our students will likely find themselves in, where they can draw on resources and use graphing and calculating tools when working towards a solution that they will have to present to others. Open book assessment can be a way to develop skills such as speaking about a mathematical concept, eliminating careless errors, working under time pressure, as well as technical skills such as creating a video portfolio. Using the resources available, students need more than knowledge and recall, but have to critically and analytically solve problems. We made use of open-ended questions (Tallman et al., 2016) that require students to give explanations for their choices. As these types of questions typically take longer than recall or procedure questions, we agreed that giving students practice opportunities in low-stakes assessments was essential before using different but similarly structured questions in assessments contributing the most towards the final course results.

We give a typical example of such questions as follows:

For each of (i), (ii), and (iii), draw an example of a continuous function that has

(i) No absolute minimum;
(ii) One local minimum, where the first derivative is undefined;
(iii) Two local maxima, of which one is also an absolute maximum.

Further examples for open book assessment in mathematics are given by Iannone (2020).

The timings of our open book assessments were sometimes constrained by our institutions. Our first high-stakes online assessments allowed students to access the questions for up to 6 hours in a 24-hour interval. To reduce potential unfairness from early test takers sharing questions and their answers with other students, we moved to synchronous assessments, typically in 3-hour intervals. We noted that the attraction towards cheating could be reduced by allowing multiple attempts at assessments and by carrying out oral validity checks where a randomly chosen subset of the class have to explain their thought processes. Something valuable we gained from the CoP conversations was the support and the determination to remain firm about the recognized advantages of open book testing and seek to maintain it going forward even during face-to-face classes.

We encourage students to **verbalize** their understanding and their processes, thereby allowing us to assess their proficiency but also as a way for them to continue the learning process. In the online context, this can take the form of a Zoom call (individually or in a group), possibly using a document camera, or video or voice recordings. With a no-video low-technology policy, we encouraged students to participate in text-based online discussion forums by asking and answering questions, posting solutions, and suggesting methods of interpreting questions. The potentially overwhelming time burden linked to verbalized assessment can be minimized by using peer reviews of student's videos, group presentations, and graded participation in discussion forums to maximize the chance of students answering questions from their peers, reducing the input from the tutor or lecturer to checking rather than responding to each question. When students' work is shared and they are given an opportunity to discuss their solutions, the class community benefits by identifying misconceptions or errors and learning different ways of solving problems sourced from each student's own network of resources.

Using a concept of **self-directed learning** (Mentz & Oosthuizen, 2016), our students were encouraged to take ownership of their own learning. In assessment tasks, they were encouraged to engage in logical thinking, critical thinking, and reasoning skills to improve on their mathematical understanding, and to attempt to learn beyond the curriculum. Necessary for successful online learning is the fulfilment of the expectation for students to take more responsibility for their learning, which requires student self-efficacy and effort regulation (Shea & Bidjerano, 2010). 'Test wrappers' (Hodges et al., 2020) are a way to get students to develop metacognitive awareness through reflecting on the types of errors made in an assessment, their preparation strategies and what steps to take when stuck. Extra emphasis should be given to a test wrapper following the first assessment given to first-year students to help them adjust to the expectations of their new education system.

Assessment holds the potential to further develop **Higher Order Thinking Skills (HOTS)** developed during learning, namely logical thinking, critical thinking, and reasoning skills (Lewis & Smith, 1993). By creating formative assessment moments during a course, students become familiar with the modes of assessment that create opportunities for the development of HOTS, in contrast to the model of a single summative assessment moment at the end of a course. Examples of HOTS-developing assessments are open-ended questions that ask students to find and explain an error in a sample of work; show-and-tell presentations of answers; and questions that require constructing, reconstructing, and changing strategies while problem solving.

Mastery learning approaches are teaching and learning strategies based on the premise that 'virtually all students can and will, in time, learn' (Anderson, 1975, p. 4). Reviews and meta analyses of mastery learning have shown positive effects for teaching, learning, and assessment (Groen et al., 2015; McGaghie, 2015). In our courses, a mastery learning approach was employed using the online assessment tools WebAssign and Grasple, providing assessment for learning through quick links to relevant textbook sections. The advantage in using an online assessment tool was access to a large database from which we could carefully choose questions to encourage students to understand and not just memorize concepts, with randomized questions, multiple opportunities for practice, scaffolded tutorial problems, video, graphical and written feedback, immediately available. The data analytics provided by these tools allowed us to quickly identify topics and specific questions to discuss in classes. While the tools allowed us to check on students' progress, this was an area with which we would like administrative support, especially in larger classes.

4. Sustaining the CoP

Our CoP initially formed through our shared experiences as female academics teaching calculus students and our past Delta conference interactions. Two key ingredients held our CoP together and allowed us to learn from one another despite our different circumstances. The first is our shared student-centred values that underpin our decisions related to teaching and assessing during the pandemic. We believe that our students will be more motivated to take charge of their learning because they see our interest and care. Second, the foundation for our interactions was trust, which provided the psychological safety that made us comfortable with being questioned or criticized.

The benefits to our teaching practices as a result of our CoP interactions encouraged us to continue our regular meetings. Our CoP meetings allow for reflections and thinking about how to challenge the system and adapt it for better outcomes, experiences, and growth. Change is possible but slow due to institutional structures. The periodic boosts we felt from our CoP meetings re-energized our teaching efforts.

Open book assessment may challenge the way learning takes place in mathematics, moving from rote, surface learning to deep understanding. This will require moving out of comfort zones and away from our question banks stored over years to a new mindset for the design of assessment that encourages such learning. It will require us as lecturers to develop new skills – we had no open book assessments when we were students. The cultural shift towards open book assessment may be slow and stifled by institutional constraints, including the fear of moving out of a comfort zone. We should expect that not all our colleagues will be as eager as us to change their assessment practices.

5. Takeaway

The Covid-19 pandemic was the catalyst that led to this group's realization that there is much value in working as a community of practice to voice concerns, ideas, and experiences regarding our teaching. Our fortnightly meetings directed us to reading and writing homework related to teaching topics we discussed. Working together on a focused task, such as writing this paper, helped to strengthen our commitment to each other and protect our meeting times from other work and life demands, as far as possible. Above all, our CoP boosted our morale. By reflecting on our improvements and challenges, we were reminded of the bigger challenge of how to make our teaching relevant to a changing world.

Disclosure statement

No potential conflict of interest was reported by the authors.

Author statement

The first author was lead author; the other three members of the community of practice contributed equally and are listed alphabetically.

ORCID

Anita L. Campbell https://orcid.org/0000-0003-4782-7323
Tracy S. Craig http://orcid.org/0000-0002-3653-2995
Batseba Mofolo-Mbokane http://orcid.org/0000-0002-3443-2640
Pragashni Padayachee http://orcid.org/0000-0001-9966-8171

References

Anderson, L. W. (1975, March 27–29). *Major assumptions of mastery learning*. Annual Meeting of the Southeast Psychological Association, Atlanta, Georgia. ERIC Online. http://files.eric.ed.gov/fulltext/ED150172.pdf.

Cnop, I., & Grandsard, F. (1994). An open-book exam for non-mathematics majors. *International Journal of Mathematical Education in Science and Technology*, *25*(1), 125–130. https://doi.org/10.1080/0020739940250116

Croasdale, R. (1973). Open-book examinations in operational research (a personal view). *International Journal of Mathematical Education in Science and Technology*, *4*(2), 133–135. https://doi.org/10.1080/0020739730040207

Groen, L., Coupland, M., Langtry, T., Memar, J., Moore, B., & Stanley, J. (2015). The mathematics problem and mastery learning for first-year, undergraduate STEM students. *International Journal of Learning, Teaching and Educational Research*, *11*(1), 141–160. https://www.ijlter.org/index.php/ijlter/article/view/303.

Hodges, L. C., Beall, L. C., Anderson, E. C., Carpenter, T. S., Cui, L., Feeser, E., Gierasch, T., Nanes, K. M., Perks, H. M., & Wagner, C. (2020). Effect of exam wrappers on student achievement in multiple, large STEM courses. *Journal of College Science Teaching*, *50*(1), 69–79. https://www.nsta.org/journal-college-science-teaching/journal-college-science-teaching-septemberoctober-2020/effect-exam.

Iannone, P. (2020). Assessing mathematics at university: Covid-and beyond. *Newsletter of the London Mathematical Society*, *490*, 34–40. https://www.lms.ac.uk/sites/lms.ac.uk/files/files/NLMS_490_for%20web2_1.pdf

Iannone, P., & Simpson, A. (2021). How we assess mathematics degrees: The summative assessment diet a decade on. *Teaching Mathematics and its Applications: An International Journal of the IMA*, *hrab007*. https://doi.org/10.1093/teamat/hrab007

Lave, J., & Wenger, E. (1991). *Situated learning: Legitimate peripheral participation*. Cambridge University Press.

Lewis, A., & Smith, D. (1993). Defining higher order thinking. *Theory Into Practice*, *32*(3), 131–137. https://doi.org/10.1080/00405849309543588

McGaghie, W. C. (2015). Mastery learning: It is time for medical education to join the 21st century. *Academic Medicine*, *90*(11), 1438–1441. https://doi.org/10.1097/ACM.0000000000000911

Mentz, E., & Oosthuizen, I. (2016). *Self-directed learning research*. AOSIS.

Shea, P., & Bidjerano, T. (2010). Learning presence: Towards a theory of self-efficacy, self-regulation, and the development of a communities of inquiry in online and blended learning environments. *Computers and Education*, *55*(4), 1721–1731. https://doi.org/10.1016/j.compedu.2010.07.017

Tallman, M. A., Carlson, M. P., Bressoud, D. M., & Pearson, M. (2016). A characterization of calculus I final exams in US colleges and universities. *International Journal of Research in Undergraduate Mathematics Education*, *2*(1), 105–133. https://doi.org/10.1007/s40753-015-0023-9

Thomson, K. E., & Trigwell, K. R. (2018). The role of informal conversations in developing university teaching? *Studies in Higher Education*, *43*(9), 1536–1547. https://doi.org/10.1080/03075079.2016.1265498

Wenger, E. (2000). Communities of practice and social learning systems. *Organisation*, *7*(2), 225–246. https://doi.org/10.1177/135050840072002. http://www.psycholosphere.com/Communities of Practice & Social Learning Systems by Wenger 2000.pdf.

Wenger-Trayner, E., & Wenger-Trayner, B. (2015). Communities of practice: A brief introduction. *Wenger-Trayner*. https://wenger-trayner.com/wp-content/uploads/2015/04/07-Brief-introduction-to-communities-of-practice.pdf.

Index

Page numbers in **bold** refer to tables and those in *italic* refer to figures.